Introduction to Early American Masonry

Stone, Brick, Mortar and Plaster

HARLEY J. McKEE, F.A.I.A.

National Trust/Columbia University Series on the Technology of Early American Building

1

Published by
NATIONAL TRUST FOR HISTORIC PRESERVATION
and
COLUMBIA UNIVERSITY

The Preservation Press
National Trust for Historic Preservation
1785 Massachusetts Avenue, N.W.
Washington, D.C. 20036

The National Trust for Historic Preservation is the only private, nonprofit organization chartered by Congress to encourage public participation in the preservation of sites, buildings and objects significant in United States history and culture. Trust support is provided by membership dues, endowment funds, contributions and grants from federal agencies, including the U.S. Department of the Interior, Heritage Conservation and Recreation Service, under provisions of the National Historic Preservation Act of 1966.

Fourth printing, 1980

Library of Congress catalog card number: 73-84522
ISBN 0-89133-006-2

FOREWORD

Although American architectural historiography has seen an unprecedented flowering in recent decades, certain areas are still neglected by the professional historian. One of these is architectural technology, especially that of the folk and vernacular aspects of building construction. The conventional knowledge of the field was traditionally transmitted from one generation of craftsmen to the next, largely through apprenticeship. Much of it was never committed to paper and all of it until recently escaped the attention of the history book. Even today, when the subject is beginning to attract the attention of scholars and specialists, there is an absolute lack of organized teaching materials on the subject. Published work is disparate, incomplete and scattered throughout innumerable periodicals in many disciplines. This is a natural reflection of the fact that current research in the field is being carried on by many specialists in many scattered institutions.

With the growth of the historic preservation movement in this country and the variety of new types of project being initiated, information on the technological (as opposed to the purely aesthetic or historical) facts becomes increasingly important.

In an effort to provide such information, the Graduate Program in Restoration and Preservation of Historic Architecture at Columbia University initiated a new course in the technology of early American building in 1968, under the general direction of Professor Charles E. Peterson, F.A.I.A., F.R.S.A. As an extension of this course, a series of lectures covering various aspects of the subject in a more detailed fashion was commissioned in 1972. Leading specialists were asked to prepare lectures on preindustrial structural theory and construction techniques, giving special attention to local response to regional environments.

It is, of course, obvious that many early American architectural practices derive ultimately from European experience. As Professor McKee points out in this monograph, American design concepts, tools and terminology were derived from those of western Europe. But because of the special conditions in the New World, all these inherited *metiers* were subject to immediate and continuous new pressures. On the one hand, such factors as more rigorous climates and abundant supplies of timber and ores led to the development of skeletal structures, first in wood, then in metal. On the other hand, the chronic shortage of skilled craftsmen gave impetus to the rationalization and mechanization of the building field; this is the fundamental technological process that underlies all the stylistic distinctions of 18th and 19th-century American architecture—the central phenomenon so consistently overlooked by conventional art and architectural historiography.

Professor McKee's monograph, which is a summary of his lectures on American masonry at Columbia University, is the first in the National Trust/Columbia University Series on the Technology of Early American Building. Others from the lecture series will be published in due course and the collection may ultimately be combined to form a basic text on early American building technology. The lectures themselves would not have been possible without the generous and ongoing support of the Edgar J. Kaufmann Charitable Foundation and the J. M. Kaplan Fund.

James Marston Fitch
Director, Graduate Program
Restoration and Preservation of Historic Architecture
Columbia University

James C. Massey
Director
Department of Historic Properties
National Trust for Historic Preservation
July 1973

PREFACE

This book contains material from short lectures given by the author at the School of Architecture at Columbia University from 1967 to 1972, and a series of formal lectures given early in 1973. The lectures were part of a seminar on the Technology of Early American Building in the Graduate Program in Restoration and Preservation of Historic Architecture, developed by Professors James Marston Fitch and Charles E. Peterson. A short essay on early American masonry materials was published by the author in 1971; the present version has been revised and expanded.

Those interested in learning about early masonry in the United States have had to visit many sites and scan many pages of old books to extract scanty bits of documentary information. This book provides a summary of the data the author has collected. It makes a basic introduction to the subject readily available to students of historic preservation, providing a point of departure for more intensive study of particular topics and localities.

Notes on materials, tools and practices from earlier times and other countries are included. The knowledge and skills of early American builders were derived in large measure from European sources, and American achievements cannot be placed in proper perspective unless they are compared with older methods. A work process and the tools associated with it cannot be fully understood without some acquaintance with their origins and development, yet knowledge about materials and methods of working them is so widely diffused that it is often impossible to discover the source of a particular method. It is necessary, therefore, to examine European techniques that are similar to those employed in America at a later date. An early American craftsman's methods were closer to those of his medieval European or ancient Roman counterpart than to modern ones.

Professional architects and engineers were important in late 18th and 19th-century American building (the two professions were not clearly differentiated until about 1870). To these men, who were acquainted with the theories and principles of building in Europe as well as practical matters, belongs much of the credit for the introduction of improved methods.

A building is a system in which all of the materials are interrelated. Although each material must be studied separately, one must not lose sight of the system as a whole. The hardness of brick or stone is related to the thickness of a wall, the mortar, the bond and the quality of workmanship; all are important in determining the load-carrying capacity, resistance to weather and appearance. Production of a specific building material may be only one part of a larger business enterprise. The technical processing of one material may be similar to the processing of another, or the production of one may not be economical without the other. Thus, bricks and tiles are often made by the same company; limestone may be quarried for use in the refining of iron ore, for building stone, for the manufacture of lime and cement or for fertilizer and gravel.

Several masonry materials are not included in this book: slate, unbaked brick, tile, concrete and artificial stone. Primary emphasis is placed on the materials, sources, tools and processes employed; little attention is given to architectural effects and social circumstances. Uncommon techniques are introduced along with common ones. The great majority of masonry structures in the colonies and in early American settlements were unpretentious, requiring for their construction only simple tools and methods. The masonry work found in monumental public buildings and fashionable mansions, although these structures were relatively few in number, is of such significance that it is discussed at considerable length in this book.

The year 1860 was chosen as the terminal date for the discussion of techniques. However, some later examples are included to avoid breaking off certain developments at an awkward point. For the most part, this book is devoted to manual work. The introduction of machinery took place gradually; manpower, horsepower and waterpower were applied to machines before steam engines assumed the burden of providing energy. But in small establishments, hand processes continued to be used well into the 20th century. The study of these processes provides us with helpful insight into the traditions they represent.

The author is indebted to several persons with whom he has exchanged information over the years, especially to Charles E. Peterson for some key references. Many topics in this book deserve more extended treatment than is given here. Readers wishing to pursue specific topics may find sources for further study in the footnotes at the end of each chapter and in the appendix.

H.J.M.

Syracuse, New York
February 1973

CONTENTS

SOURCE NOTES

The following three drawings were reproduced from the sources cited:

— Bricklayer's Hoist on p. 50 and Chamberlain's Brick-Machine on p. 45 from Edward H. Knight, *Knight's American Mechanical Dictionary* (New York, Hurd and Houghton, 1877), vol. 1, p. 373, figs. 907-908.

— Terra-cotta Window on p. 55 from *The Brickbuilder*, vol. 5, no. 12 (December 1896), p. 228.

All other drawings were made by the author. The following drawings were patterned after illustrations from the sources cited:

— Iron Wedges on p. 16 Nos. 2-6 from Charles Singer et al., eds., *A History of Technology*, 5 vols. (London: Oxford University Press, 1954, 1957), vol. II, pp. 15, 36.

— Bunker Hill Quarry Machines on p. 19 from Arthur W. Brayley, *History of the Granite Industry of New England* (Boston: National Association of Granite Industries of the United States, 1913), vol. 1, p. 64.

— Stone Boat on p. 18 from Edward H. Knight, op. cit., vol. 3, p. 2392, fig. 5844.

— Drills on p. 17. No. 1 from Norman Davey, *A History of Building Materials* (London: Phoenix House, 1961), p. 227, fig. 123. No. 3 from Giorgio Vasari, *Vasari on Technique* (New York: Dover, 1960), p. 48, fig. 2.

— Picks on p. 23. Nos. 1-2 from Norman Davey, *op. cit.*, p. 227, fig. 123. No. 3 from Giorgio Vasari, *op. cit.*, p. 48, fig. 2.

— Axes and Hammers on p. 22. No. 1 from Frederick P. Spalding, *Masonry Structures* (New York: John Wiley & Sons, 1921), p. 63, fig. 5. nos. 4-5 from Norman Davey, *op. cit.*, p. 227, fig. 123. Nos. 6-8 from F. E. Kidder, 9th ed., *Building Construction and Superintendence* (New York: Comstock, 1910), p. 269, fig. 104; p. 270, fig. 108.

— Chisels on p. 29. No. 3 from Frederick P. Spalding, *op. cit.*, p. 65, fig. 13. No. 4, 6-8 from F. E. Kidder, *op. cit.*, p. 270, fig. 109.

— Points on p. 24. No. 1 from Norman Davey, *op. cit.*, p. 227, fig. 123. No. 2 from Giorgio Vasari, *op. cit.*, p. 48, fig. 2. No. 3 from F. E. Kidder, *op. cit.*, p. 270, fig. 109.

— Pitching Chisel or Pitcher on p. 26. from F. E. Kidder, *op. cit.*, p. 270, fig. 109; p. 271, fig. 110.

— Stoneworking on p. 31 from Denis Diderot, *A Diderot Pictorial Encyclopedia of Trades and Industry*, Charles Coulston Gillespie, ed. (New York: Dover, 1959), p. 282, pl. 276, vol. 2.

— Bricklayers' Tools on p. 46 from Edward Lomax and Thomas Gunyon, eds., *Encyclopedia of Architecture, a New and Improved Edition of Nicholson's Dictionary* (New York: Johnson, Fry & Co., n.d.).

— Lime Kilns on p. 62. No. 4 from Eli Bowen, *The Pictorial Sketch-Book of Pennsylvania* (Philadelphia: Willis P. Hazard, 1852), p. 45.

All photographs except those specifically credited to others were taken by the author.

I. STONE

INTRODUCTION

Stone has been used as a building material for the finest buildings in many lands, buildings that endured where structures of other materials decayed. Because they are available for the architectural historian to study firsthand, old stone buildings have received a disproportionate share of attention, as compared with buildings constructed of less-durable materials. The technology employed in their construction, however, is but sparingly mentioned in publications on American architecture.

A relatively small number of the historic buildings in the United States were built with stone walls. Among those that were, the use of arches and vaults was uncommon. Only the most expensive public and private structures were elaborately finished in early times, but as more people became prosperous there was an increased demand for marble chimney pieces, stone steps and sills and other such elements. Even simple buildings in the colonies had stone foundations and sometimes had stone chimneys; however, this material was usually given only a rough finish. The stonework of most houses and smaller public buildings was informal in character, revealing a variety of sizes, shapes and colors characteristic of natural material. Contemporary architects often admire this informality and consider it a virtue of early American architecture. Yet the more laborious craftsmanship of monumental stone buildings also has its place in the study of early American architecture, even though it is less frequently encountered.

Nature of Stone. Stone is a brittle material, better able to support heavy weights than to be subjected to the stress of bending. It is most effectively used in walls and piers but can be used in beams of limited span. Large heavy stones in massive groups can resist active forces because of their inertia and because only great forces can overturn them. Certain kinds of stone are better able to resist wear, exposure to the weather and soiling than others. Early builders who possessed the resources to procure and transport material from distant quarries could select the specific kinds they desired, especially for finish, but most builders had to be contented with those types of stone that could be obtained near a building site. Local varieties were always the most commonly used; in areas where good building stone was found, stone construction flourished and appropriate technology developed. Modern transportation facilities greatly diminished the contrast between building methods in different regions, but differences in materials and technology are readily apparent in the study of historic structures.

The term *masonry* originally applied only to stonework, but in the United States today the word is widely used to refer to works in brick, concrete and even earth. *Rock* refers in general to the materials that form the outer part of the earth; *stone* refers to the hard portions and to the pieces that are used in construction. A *quarry* is an open excavation from which stone is removed.

Formation of Rocks. Rocks are generally classified according to their manner of formation. The formative processes, however, are complex, and not always clearly distinguishable. *Igneous* rock is formed by volcanic action. Primary material *(magma)* is plastic at high temperatures and pressures. When forced up toward the surface, where temperature and pressure are lower, this material may crystallize into *granite* (or other igneous rock, according to which fractions of the magma solidify). Some granites have large crystals—the result of slow cooling—and others have smaller crystals. If magma is forced through the earth's crust quickly, as in a volcanic eruption, dissolved gases are released, expanding the material into porous scoria, pumice or volcanic ash.

Igneous rocks disintegrate on the surface of the earth. Those exposed to heating by the sun and cooling at night, to rain and to the various forces of erosion are eventually broken up into smaller pieces and grains. These are transported by streams or wind and deposited in river valleys, lakes or oceans. *Sedimentary* rocks of varied composition are formed by the accumulation of these sediments, to which may be added the shells of marine organisms and substances precipitated out of seawater, such as calcium carbonate and magnesium carbonate. The circumstances under which sediments accumulate affect the final composition of the rock: Gray or blue clays result when sediment is continually underwater; reddish clays result when sediment is alternately exposed to water and air, allowing some of its iron content to oxidize. The weight of succeeding sediment helps to consolidate the rock below; particles of sedimentary rock may also be held together by cementitious matter, such as calcium carbonate, silica, iron oxide or clay.

The most common sedimentary building stones are *sandstone*, in which silica predominates, and *limestone*, in which calcite predominates. *Shale*, composed of mud or clay, is sometimes used in walls, but it is generally considered an inferior building material. *Tufa* is a soft stone formed from volcanic ash. *Conglomerate* consists of pebbles in a cementitious material; *shellrock* or coquina is composed of shells similarly held together.

Igneous and sedimentary rocks are changed into *metamorphic* rock by various processes. These processes include pressure from movements of the earth's crust, heat from intrusions of magma and the action of groundwater and substances dissolved in it. Texture may be altered by molecular rearrangement or by a change in composition; in general, crystals become more apparent than they were in the parent rock. Granite is transformed into *gneiss*, limestone into *marble* and shale into *slate*. The transformation is not always complete, and rocks of intermediate nature are common.

Material dissolved in groundwater is sometimes precipitated to form rocks. This is especially true near hot springs, where *travertine* and *pot rock* may be found. These types of rock often have a spongy or worm-eaten appearance. Stalactites and stalagmites in caves are precipitated from material carried in solution by groundwater.

Potomic breccia, also called Potomac marble and calico rock, used in a column in the old House of Representatives, U.S. Capitol, Washington, D.C. This material, which was quarried about 1816 near Great Falls of the Potomac River, was first used in construction by Benjamin H. Latrobe, an Architect of the Capitol. (William Edmund Barrett)

Any rock may be lifted by movements of the earth's crust and subjected to one or more cycles of disintegration, erosion and deposition. *Breccia*, a rock consisting of sizable chunks of marble and other stone cemented together in a variegated pattern, is an example of such "recycled" material. It is highly valued for decorative purposes, although difficult to work.

Sizes and Properties of Rocks. Rock formations in the earth's crust may include unbroken pieces of enormous size. David Dale Owen, reporting to the Smithsonian Institution in 1848, described granite in the Woodstock, Md., area. ". . . one may see a perpendicular face of nineteen feet presented to view, extending twenty, thirty, and even forty or fifty feet, without a seam or flaw, or the slightest variation in hue. A mass of forty or fifty tons weight may often be seen severed from the parent rock, by the simple but effective means of small iron wedges . . . ."[1] Other formations are found broken into pieces from several inches to several hundred feet long. Such breaks are caused by thermal changes, shock waves from earthquakes or periodic vibrations. Augustus Porter, in a letter dated January 3, 1817, thus described the common limestone rock found between Lake Erie and the Genesee River in western New York state: "The kind of rock is horizontal strata or layers of limestone, of from 6 to 24 inches thick. The horizontal joints, between these layers, are so open, that there is very little difficulty in separating the layers. The layers are separated by perpendicular cracks, dividing them into irregular and unequal slabs, of from 1 to 6 or 8 feet square."[2] Charles E. Foote, a geologist, described the natural jointing of blue sandstone in Ulster County, N.Y., in about 1900:

> The beds of stone in the Ulster quarries are divided naturally by vertical joints at right angles; one system running north and south and the other east and west. The east and west joints are known to the quarrymen as the "heads" or "headers," while those running north and south are called "side seams." These seams are five to seventy-five feet apart, and the distance determines the maximum size of the stone that can be taken from them. The layers into which a ledge is divided by horizontal seams are known as "lifts." These are split apart by means of thin wedges, driven to make a practically uniform pressure along the entire front, so as to raise the layer back to the next joint. . . .[3]

The natural divisions in rock, described previously, are related to the crystal lattice or molecular shape of the material, so that the manner in which a given stone will break depends upon the kind of crystalline structure it possesses. Instruments that have become available in recent decades enable geologists to investigate the molecular structure of minerals and correlate it with phenomena recorded in earlier times. Often, sedimentary stone will readily split along a plane in which sediment was originally deposited—its "natural bed." Many stones can be broken along a system of planes which permit rectangular, or approximately rectangular, pieces to be shaped

Natural rectangularity of rock masses. Puye cliffs, N.M.

Plasticity of stone. Originally flat, this tombstone gradually bent under the pressure of its own weight. Baltimore, Md.

with moderate effort and simple tools. Builders long ago discovered this property and learned how to take advantage of it for building purposes. The direction of the easiest cleavage is usually called the *rift;* the second easiest is called the *grain.*

In some places where the earth's crust has been subjected to great pressure, layers of rock have been bent and folded over long periods of time. Stone's plastic nature is also revealed in some locations where natural rock was under compression; after trenches were cut the stone expanded. A mass of granite 60 feet long was found to have expanded two inches at Stone Mountain, Ga.[4]

SOURCES OF BUILDING STONE IN THE UNITED STATES

Glacial Deposits. Builders have always used the most accessible stone that would satisfy their purposes. In the northeastern part of the United States—a glaciated area—the first deposits to be used were the many irregularly rounded boulders found on the surface of the ground or thinly covered by soil. Such stones from glacial deposits are of varied and sometimes quite distant origin; they are rounded, in contrast to the usually angular stones found south of glaciated regions. Granite, trap, greenstone, sandstone and gneiss boulders were once found in Connecticut.[5] Although these stones could have been broken up with a mason's hammer, the early settlers did not shape them. Field stones shaped by the forces of nature were also used in the Hudson Valley and northern New Jersey. Many of the granite boulders from Braintree, Mass., which were employed in the construction of King's Chapel in Boston in 1749, were so large that they had to be broken up and the pieces shaped into blocks.

Natural Erosion. Stones broken off from mountains by natural erosion have also been a convenient source of building material. The great Mormon Salt Lake Temple in Salt Lake City, Utah, was constructed in part of light gray granite found on the surface of Little Cottonwood Canyon. A photograph of the 1860's displayed in the Latter-Day Saints Museum in Salt Lake City shows workmen splitting enormous blocks of rock on the slope of the canyon.

Outcrops and Ledges. Early builders often took stone from outcropping formations where the rock was easily located. Such exposed material might not have been of good enough quality to exploit commercially but it could be removed by a few men and used satisfactorily for house walls and similar construction. Stone was often exposed in the beds of streams; it could be removed when the water was low. In about 1720, a house and mill were built of quartzite taken from a creek just south of Fisherville, Pa., by a Mr. Edge.[6]

Abandoned quarry that supplied sandstone for the building of the Starrucca Viaduct (1846) in Lanesboro, Pa. Massive rock is visible in the lower half of the photograph. Stevens Point, Pa.

Delaware Quarries, a sandstone deposit first worked in about 1758. It is still active. Lumberville, Pa.

Ledges of rock exposed by stream erosion, especially those found at the outer curve of a bend in a stream, provided many sites for quarrying. At Lumberville, Pa., on the Delaware River a sandstone quarry was first opened about 1758.[7] It is still in use. As early as 1665 the inhabitants of Middletown, Conn., held in common a quarry on the east side of the Connecticut River. Its use by others was forbidden.[8] Nearness to navigable water was a great advantage to all quarries.

Types of Quarrying. Three general types of quarrying can be distinguished: (1) taking stone from the surface of the earth, especially in the form of loose pieces; (2) quarrying for occasional use; and (3) commercial quarrying, with full-time crews and equipment. Commercial quarrying did not develop in America until sufficient steady demand made it profitable in competition with other enterprises. The need for gravestones, limestone for processing iron ore, stone for producing lime, and building material for public engineering works helped stimulate the beginning of a real industry.

Geographical Distribution of Sandstone, Limestone, Granite and Marble. In the eastern United States, granite and allied rocks are found along the eastern edge of the Appalachian Mountains; marble is found along a large part of the western edge.

Limestone and sandstone are more widely distributed. A simplified summary of a complex geological history indicates that for a 175-million-year period, beginning about 425 million years ago, a great

TABLE 1—EARLY SANDSTONE QUARRYING

Date	*Place*	*Quarry or Quarrier*
by 1639	Hartford, Conn.	
by 1665	Portland, Conn.	
by 1665	Middletown, Conn.	
by 1700	Newark, N. J.	
by 1757	Aquia Creek, Va.	
c. 1758	Richmond County, Va.	
c. 1758	Lumberville, Pa.	
1767	Middletown, Pa.	
1774	Seneca Creek, Md.	
1777	Danville, Pa.	
c. 1780	Chagrin Falls, Ohio	
c. 1780	Avondale, Pa.	
1796	Beaver County, Pa.	
by 1800	Hummelstown, Pa.	
1800	Brownsville, Pa.	
1800	Chapel Hill, N. C.	University Quarry
c. 1814	Buena Vista, Ohio	Joseph Moore
c. 1820	Berea, Ohio	
by 1824	Potsdam, N. Y.	
c. 1828	Reynoldsburg, Ohio	
c. 1829	Virginville, Pa.	
c. 1829	Coeymans, N. Y.	
1831	Saugerties, N. Y.	
c. 1831	Quarryville, N. Y.	
c. 1832	Waverly, Ohio	
1835	Waynesburg, Pa.	
c. 1835	City Ledge, Ohio	John Loughry
by 1838	Battle Creek, Corunna, Jacksonburg, Jonesville, Marshall, Napoleon, Mich.	
1840	Lackawaxen, Pa.	
c. 1840	Mansfield, Ohio	
by 1842	Burlington, Vt.	
1846	Stevens Point, Pa.	
c. 1847	Portsmouth, Ohio	
by 1850	Fulton, N. Y.	
by 1855	North Belleville, N. J.	
by c. 1855	South Amherst, Ohio	
c. 1860	Carson City, Nev.	

TABLE 2—EARLY LIMESTONE QUARRYING

(Not including stone for producing lime)

Date	*Place*	*Quarry or Quarrier*
1740	Harrisburg, Pa.	
by late 1700's	Northwestern New Jersey	
1793	Washington, Pa.	
1793	McConnelsburg, Pa.	
c. 1800	Columbus, Ohio	
1805	Burlington, Vt.	Levi Willard
c. 1807	Bardstown, Ky.	
by 1817	Niagara County, N. Y.	
by 1820	Syracuse, N. Y.	
by 1821	Easton, Pa.	
c. 1830	Everett, Pa.	
1831	Downingtown, Pa.	
by 1837	Auburn, N. Y	
c. 1837	Becrofts Mountain, N. Y.	
by 1838	Grosse Isle, Monguagon, Monroe, Mich.	
c. 1840	Centerville, Piqua, Sandusky, Ohio	
by 1842	Isle La Motte, Vt.	
c. 1846	Springfield, Ohio	
by 1850	Dayton, Ohio	
c. 1850	Delaware, Ohio	
by c. 1850	Chaumont, N. Y.	
1853	State College, Pa.	
Shellstone (Coquina)		
1671	Anastasia Island, Fla.	
Conglomerate (Pudding Stone)		
by 1818	Roxbury, Mass.	

TABLE 3—EARLY GRANITE QUARRYING

Date	*Place*	*Quarry or Quarrier*
1648	New London, Conn.	
by 1715	Braintree, Mass.	
c. 1780	Haverhill, N. H.	Catamount Quarries Black Hill Quarries
c. 1803	Milford, N. H.	
by 1806	Ellicott City, Md.	
1810	Chelmsford, Mass.	
by 1812	Concord, N.H.	
by 1812	Marlborough, N. H.	
by 1814	Barre, Vt.	Robert Parker
by 1816	Cecil and Baltimore Counties, Md.	
1816	Port Deposit, Md.	
1816	Fitzwilliam, N. H.	
by 1820	Hooksett, N. H.	
1823	Cape Ann, Mass.	
1824	Sandy Hook (Gloucester), Mass.	
1825	Quincy, Mass.	Bunker Hill Quarry
1825	Augusta, Me.	
c. 1825	Anisquam, Mass.	
1827	Rockport, Mass.	
c. 1832	Woodstock, Md.	
c. 1833	Raleigh, N. C.	Capitol Quarry
1836	Waldo County, Me.	
c. 1836	Spruce Head (Thomaston), Me.	
1840	Roxbury, N. H.	
1840	Barre, Vt.	Pliny Wheaton
1848	Barre, Vt.	Ira P. Harrington
1848	Bay View, Mass.	
1852	Frankfort, Me.	Mt. Waldo Granite Co.
c. 1853	Little Cottonwood Caynon, Utah	

Gneiss

1643	New London, Conn.
by 1682	Philadelphia, Pa.
1785	Swarthmore, Pa.
1792	Haddam Neck, Conn.
early 1800's	Baltimore, Md.
1837	Hallet's Cove (Long Island), N. Y.
by 1838	Salem, N. H.

TABLE 4—EARLY MARBLE QUARRYING

(Not including marble for producing lime)

Date	*Place*	*Quarry or Quarrier*
c. 1776	Montgomery County, Pa.	
c. 1780-85	Morristown, Pa.	
c. 1784	Marble Hall, Pa.	
1785	East Dorset, Vt.	Isaac Underhill
1785	Rutland, Vt.	
1788	Isle La Motte, Vt.	
c. 1790	West Stockbridge, Mass.	Boynton's Quarry
c. 1800	New Milford, Conn.	
1804	Middlebury, Vt.	E. W. Judd
c. 1808	Upper Merion Township., Pa.	
by 1809	Sing-Sing (Mt. Pleasant), N. Y.	
by 1813	Marbletown, N. Y.	
by 1814	Cockeysville, Md.	
by 1815	Dover, N. Y.	
by 1818	Gouverneur, N. Y.	
1820	East Dorset, Vt.	Freedley Quarry
by 1824	Potsdam, N. Y	
1825	Hailesborough, N. Y.	
by 1827	Frankfort, Ky.	
by 1831	Arlington, Vt.	
1834	Plymount, Vt.	
1836	Sutherland Falls, Vt.	
before 1837	New Lebanon, N. Y.	
1838	Hawkins County, Tenn.	
1838	West Rutland, Vt.	
by 1840	Downington, Pa.	
c. 1840	Pickens County, Ga.	
c. 1840	Egremont, Mass.	
before 1842	Brandon, Clarendon, Danby, Manchester, Pittsford, Shaftsbury, Swanton, Tinmouth, Wallingford, Vt.	
Breccia		
c. 1816	Loudoun County, Va., & Montgomery County, Md.	
Serpentine		
before 1842	Cavendish, Vt.	

deal of loose rock material from mountains along the present eastern coast of the United States was deposited in an area west of a line drawn roughly from Atlanta, Ga., to Albany, N.Y. (much of which was part of an inland sea). Near these mountains, sandstone was formed, while farther west, extending to the Mississippi River, thick sheets of limestone predominated. In various places, subsequent geologic actions left these deposits exposed and ready for quarrying.[9]

Tables 1-4 provide partial lists of places where quarrying was carried on from the early days of settlement in America to the middle of the 19th century and will give the reader an idea of the geographical distribution of the various kinds of building stone.[10]

QUARRYING: METHODS AND PRACTICES

Simple Methods. Stone found in relatively thin layers in level terrain could be taken out by simple methods, as in the quarrying of coquina on Anastasia Island, Fla., in 1671.

In the quarries 3 leagues from the presidio, Indian peons chopped out the dense thickets of scrub oak and palmetto, driving out the rattlesnakes and clearing the ground for the shovelers to uncover the top layer of coquina. Day after day Alonso Diaz, the quarry overseer, kept the picks and axes going, cutting deep grooves into the soft yellow stone, while with bar and wedge the peons broke loose and pried up the rough blocks—small pieces that a single man could shoulder, and tremendously heavy, waterlogged cubes 2 feet thick and twice as long that six strong men could hardly lift from the bed of sandy shell. As a layer of stone was removed, again the shovelmen came in, taking off the newly exposed bed of loose shell and uncovering yet another and deeper stratum of rock. Down and down the quarrymen went until their pits reached water and they could go no farther.[11]

A comparable method was used to extract harder stone in Ohio in about 1820.

In early days the Berea sandstone was quarried at the surface of the formation only. The layers were thin, and were broken to desired size by means of steel wedges. If the layers were not of the proper thickness, they were split. The demand for stone was small, so that primitive methods of work gave an adequate supply.

With the growth of the industry it was found necessary to work at greater depth and in thicker courses. When a given area had been stripped of glacial drift, shales or shell-rock, it was necessary to detach the area from the surrounding rock. This was done by quarrymen who cut channels or trenches with picks in two directions at right angles to each other along the borders, the width of the channels being sufficient, of course, to admit the body of the workman. Not only was this slow work, and in summer very hot, but the dust produced attacked the lungs and throat of the workmen, sometimes with fatal results. When the trenches were cut to the desired length and width, and the area of stone thus freed in two directions, drilling and blasting were resorted to, and blocks of the desired size obtained.[12]

Channeling and Terracing. Deep quarries in early America were worked as they had been in ancient times. First, a cliff face was located or a trench was cut. On the floor of the recess or trench, blocks of the desired size were outlined and then separated by channels several inches wide, cut to the desired depth. To make channels in hard stone the Egyptians employed pounding balls of dolerite. The Romans used picks or drills. A block was broken away from its bed by a row of wedges. Wooden wedges were driven tight and then wet to make them expand, or metal wedges were driven in with a sledgehammer. In the mid-19th century, wooden wedges were used to split off blocks from the face of a cliff at a limestone quarry near Fisherville, Pa. Several hours after the wedges were wet, the blocks split loose.

As quarrying progressed, a series of terraces was formed, each terrace several blocks in width. Eventually, the lower terraces were removed, leaving a nearly vertical face marked by horizontal traces along the cliff. The ancient Athenian quarries on Mount Pentelicon still display this kind of marking.

To avoid hauling extra weight, blocks were cut to the approximate shape and size desired for building purposes before being taken from the quarry. Stones were occasionally stored at the quarry site to test their ability to withstand the weather. Vitruvius recommends this in *The Ten Books on Architecture:* "Let the

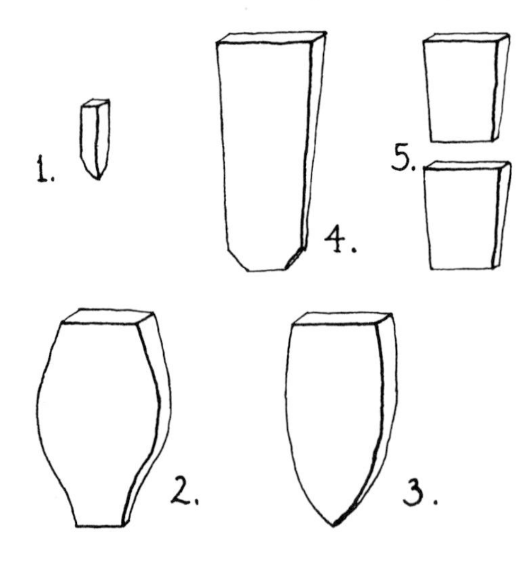

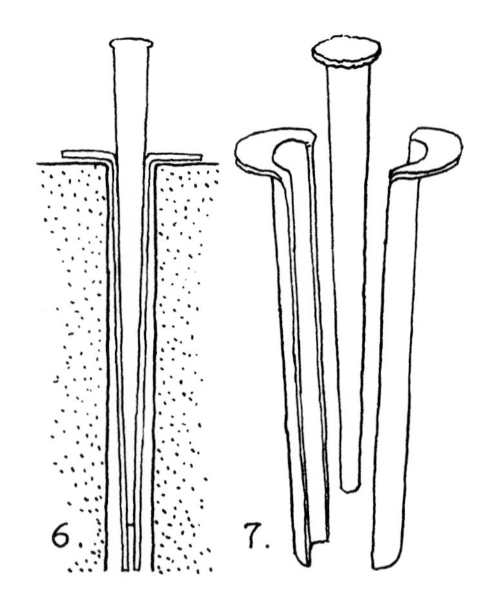

IRON WEDGES USED FOR SPLITTING STONE.

1. 2¾ inches long, ½ inch square. Mercer Museum, Doylestown, Pa.

2-3. Flat wedges.

4-5. Modern English wedges with slats.

6. Square plug and feathers.

7. Round plug and feathers.

stone be taken from the quarry two years before building is to begin, and not in winter but in summer. Then let it lie exposed in an open place. Such stone as has been damaged by the two years of exposure should be used in the foundations. The rest, which remains unhurt, has passed the test of nature and will endure in those parts of the building which are above ground."[13]

Rocks near the surface of the ground, particularly in humid climates, are usually weathered. Material for monumental buildings was obtained by removing the 15 or 20 feet of rock on the surface and cutting the better material beneath. The weathered stone, called *rag* in England, often has actual or incipient fissures that make it unsuitable for cutting into regular blocks; it was sometimes employed in rubble walls. Some of the first marble quarried in New England was located near intrusions of granite. It looked good and was easily split off and divided into useful blocks. Upon exposure after being laid in walls, however, the incipient cracks caused by heat from the nearby granite opened up. After this, quarriers were more careful to use only material that they were certain was sound, even though it was more difficult to work.

Blasting, Drilling and Splitting. Blasting was used to a certain extent in early American quarrying. This practice was confined largely to clearing away the upper unwanted layers of rock and to extracting stones for building roads and burning lime. Blasting tends to cause incipient cracks in rock, so it was usually avoided near materials of high quality.

By the middle of the 18th century, *drills* or *jumpers* were employed for cutting holes in rock in Philadelphia. However, these tools do not appear to have been used in New England until the end of the century. A drill is a long piece of iron with a chisel-like end; it was struck with a sledgehammer by one workman and held by another, who rotated it after each blow. Smaller drills could be operated by one workman. John Bartram of Pennsylvania described the process of drilling and splitting in a letter written January 24, 1757:

> I have split rocks 17 feet long and built four houses of hewn stone, split out of the rocks with my own hands. My method is to bore the rock about 6 inches deep, having drawn a line from one end to the other, in which I bore holes about a foot asunder, more or less, according to the freeness of the rock; if it be 3 or 4 or 5 feet thick, 10, 12, or 16 inches deep. The hole should be an inch and a quarter in diameter of the rock be 2 feet thick, but if it be 5 or 6 feet thick the holes should be an inch and three quarters diameter. There must be provided twice as many iron wedges as holes, and one-half of them must be fully as long as the hole is deep and made round at one end, just fit to drop into the hole, and the other half may be made a little longer and thicker one way, and blunt pointed. All the holes must have their wedges drove together, one after another, gently, that they may strain all alike. You may hear by their ringing when they strain well. Then with the sharp edge of the sledge strike hard on the rock in the line between every wedge, which will crack the rock; then drive the wedges again. It generally opens in a few minutes after the wedges are drove tight. Then, with an iron bar or long levers, raise them up and lay the two pieces flat and bore and split them in what shape and dimensions you please.[14]

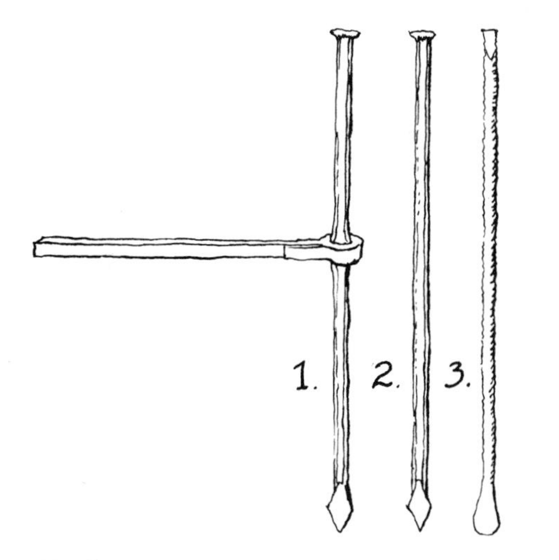

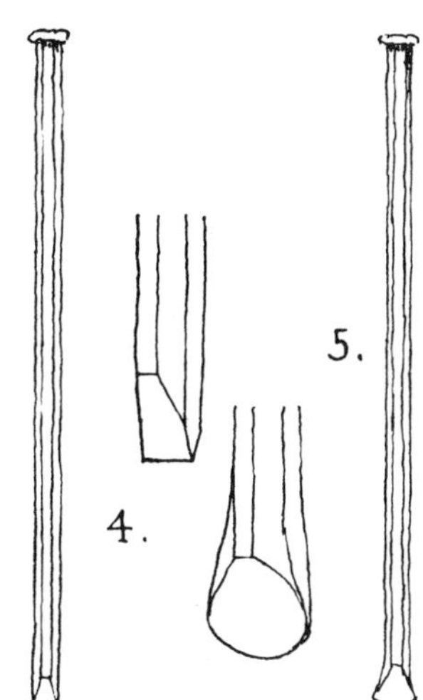

DRILLS.
1. 19th century.
2. 19th century.
3. Italian, 16th century.
4. 23 inches long, 1 inch octagonal. Mercer Museum, Doylestown, Pa.
5. 31½ inches long, 1 inch octagonal, 1¾ inch cutting edge. Mercer Museum.

Albert Hager, who visited a marble quarry in Dorset, Vt., in 1858 described the use of drills to cut channels one or two inches wide around blocks, which were then split off with wedges. "...it costs twenty-eight cents a foot to get these channels cut. A good workman will cut from five to ten feet—that is a groove one foot deep and from five to ten feet in length—per diem, which yields a daily income to the workman of $1.40 to $2.80."[15] By about 1880 steam-powered channeling machines were being employed in the quarries of New England.

Lifting and Transporting. Handling large stones, which weigh from 140 to 180 pounds per cubic foot, is difficult. The great masses of stone split off in the quarry were usually subdivided on the spot, to the approximate size needed for building. If the subdivided pieces were too heavy to be turned or lifted by a few men, levers were used to manipulate the material onto a sledge, stone boat or cart. The sledges and carts carried the material along a ramp to the surface. Large stones were sometimes moved on wooden rollers. Cranes, consisting of booms and vertical masts held by stays, could handle fairly heavy loads with the mechanical advantage afforded by windlass and compound pulleys. They could easily be taken down and moved. When steam engines became available to provide power, heavy loads were lifted more rapidly; in the late 19th century, steam-powered hoists were common in American quarries. By that time, deep quarries were often equipped with strong cables stretched between towers, along which carriages were moved, carrying hoisting tackle. The tackle was supplied with power from a stationary engine in a shed at the rim of the quarry.

The methods used at the Bunker Hill quarry, in Quincy, Mass., provide good examples of advanced 19th-century quarrying techniques.[16] Opened in 1825 to supply granite for the Bunker Hill Monument in Charlestown, this quarry was operated by the monument commissioners, under the management of Solomon Willard, the architect, and Gridley Bryant, the master mason. Later it was worked commercially.

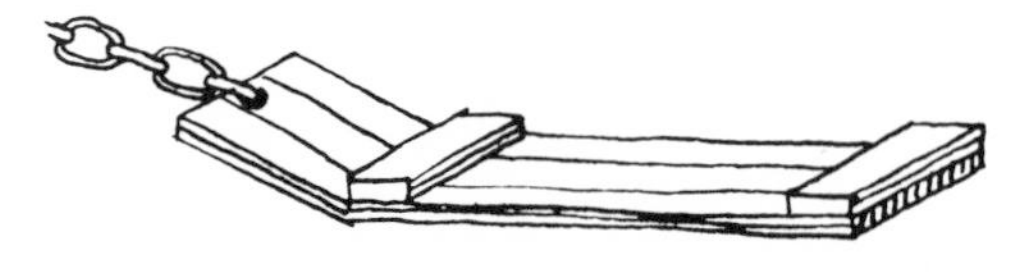

STONE BOAT. Used for hauling stones.

Willard and Bryant were capable and versatile men—energetic, inventive and practical. They were faced with many difficult problems in operating the quarry, not the least of which was handling and transporting material in large pieces from the quarry to the building site, which was 12 miles away. A railway was constructed to carry blocks of stone to a landing on Gulliver's Creek, near the Neponset River. Grades varied from 13 to 36 feet to the mile; at the quarry, an elevation of 84 feet had to be surmounted. For this, an inclined plane 315 feet long, at an angle of 15 degrees, was built. An endless chain drew cars up and down the ramp. Horses drew cars along the railway and furnished power for the inclined plane. The track continued at different grades around the quarry proper.

Each carriage or "truck" had two pairs of wheels, connected so as to allow swiveling; each was capable of carrying a six-ton load. Larger loads were managed by joining trucks together: Two could carry 10 tons. Four trucks were attached to one another to form a 16-wheel car to transport columns for the Old Court House in Boston in 1836. Each of the columns weighed 64 tons in the rough. Stones for the Bunker Hill Monument were drawn by horses along the railway to the landing. There they were put on flat-boats and towed by a steamboat to Deven's Wharf in Charlestown. From there they were taken to the building site by teams of horses. So many stones were damaged by the transfers that after the first few courses of the monument had been laid, stones were teamed directly from the quarry. The railway continued to be used for about 25 years to handle other shipments, however.

Willard and Bryant also devised an improved apparatus for turning and lifting stones. "The lifting jack has been found to be a useful machine for turning heavy blocks of stone. It is a compact and powerful machine, calculated for hard service. . . . It consists of a rack, and one or more wheels and pinions, according to the power required. . . .the whole was made of the best of wrought iron and cast steel, except the boxes, which were of bronze or composition. The rack and wheels were of wrought iron, and the pinions of cast steel."[17]

Other devices described by Willard were used for pulling and hoisting; they could be operated by one man.

The pulling jack is constructed much like that for lifting, but is always in a horizontal position. The crank pinion is two or three feet, and turned by four arms about three feet long. The rack has a claw at the end to receive a chain, which may be led to places inaccessible and dangerous for using the common jack. It is a powerful and convenient purchase for carting and hauling out heavy blocks of stone.

The power of the one used is about 10 tons . . . . This machine was contrived and first used at the Bunker Hill quarry.

The hoisting apparatus . . . is calculated for raising weight too heavy for shears or derricks, and has been found convenient for loading any stone from 5 to 50, or even 60 tons in weight. A horse, or timber frame, is set over the stone to be raised, supporting a screw and nut. A chain from the weight leads to a shackle, which is connected with the screw. The nut is then turned round by long arms, and the weight raised to a proper height for the carriage to pass under it, and when properly adjusted the weight is lowered to its bearings.

For blocks of granite of great length . . . two horses and screws were used.[18]

For the construction of the Erie Canal, beginning in 1817, stone was transported on sleds during the winter. This was especially advantageous in some of the marshy country through which the canal passed. There, a heavy load could not be supported unless the ground was frozen. A report of the canal commissioners dated January 25, 1819, expressed the hope that there would be five weeks of good sleighing that winter.

In New England, heavy stones were transported overland by oxen. Winter was the best season for doing this because the animals were not needed for farm work then. In 1836, the granite columns for the old Court House in Boston were transported in the following manner:

A team of 65 yoke of oxen and 12 horses were required to draw them, and the other monster monoliths from the railway in Quincy to Boston, a distance of over ten miles. While the mighty stones were being loaded by jack screws in the wagon the whole country for miles around had been scoured for oxen to haul it. Oxen were the more accessible beasts of burden in that section, as in all other rural districts. . . . At a specified hour drivers each with a yoke of oxen, and men driving two horses, gathered around the wagon. There were two chains fastened to the cart, and to each of these two oxen were attached, making four abreast. The horses were hitched tandem fashion at the head of the procession, and each driver had his own yoke of oxen to look after. . . . The long cavalcade attracted much attention along the route. . . . The cost of transporting each column was $100.[19]

Quarries on or near navigable water enjoyed an advantage over others. Stone was moved short distances on rafts and shipped considerable distances by boat. Material for trim, monolithic columns and lintels was sent from northern ports to Charleston, Savannah and New Orleans, as well as to marble factories in New York and Philadelphia. Some trim was imported from England in the 18th century as, for example, that used at Westover Plantation in Virginia. Cobbles carried on oceangoing ships as ballast were used to pave some streets in Charleston, S.C.

As a rule, however, the importation of large amounts of stone was prohibitively expensive. The committee for King's Chapel in Boston wrote to Ralph Allen of Bath, England, in 1750 to ask if he would donate freestone from his celebrated quarry for the chapel. ". . . as there is no such thing as free Stone in these Parts, they have begun to make Use of such as the Country affords, which is so hard and course [sic.] a Nature that it is incapable of being wrot into any thing Ornamental, such as the Jammes of Doors and Windows, Pediments, Capitals, and the like, nor if such Stone could be found is there any Workman capable to do it."[20] Allen agreed to give all of the stone that was needed, delivered it to his wharf and estimated the cost of sending a master and six workmen from Bristol to Boston to work and install the stone at nearly 1,200

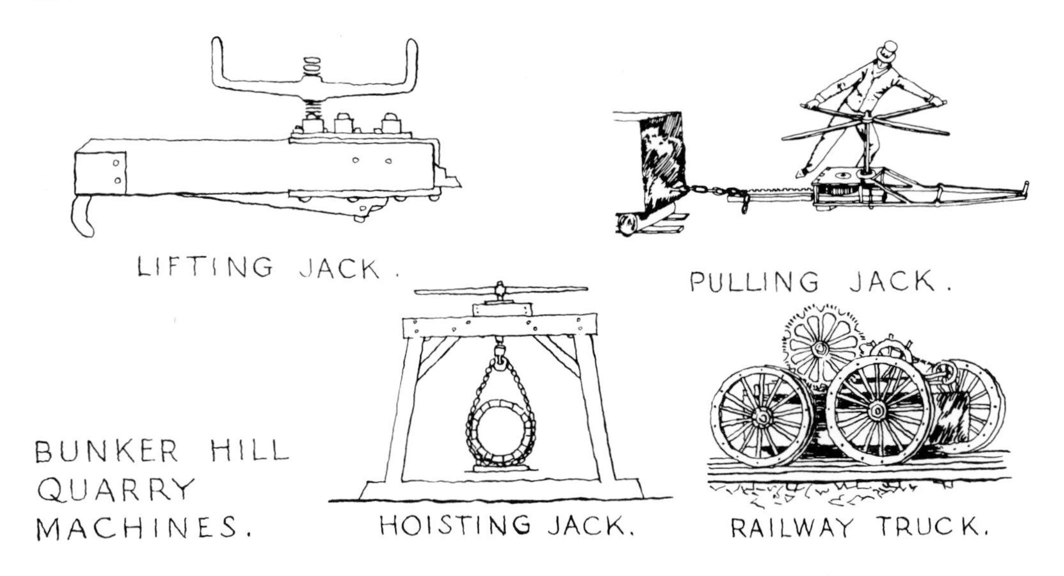

BUNKER HILL QUARRY MACHINES.

pounds sterling. This amount was beyond the resources of the church, so Allen's offer of the material was declined.

Stone Measure. Building stone was measured in different ways. Contracts for the Erie Canal, in 1817, called for measurement in *cords*:

The stone shall be delivered and corded near the site of the said aqueduct, and convenient for its construction, in such pile or piles, and place or places, as either of the said commissioners, or of the engineers in their employment, shall direct at any time before their delivery. The stone shall be thick, large, solid, rectilinear, well shaped and well faced, and in all other respects completely adapted to the construction of a substantial and permanent aqueduct. One-twentieth part of the whole quantity shall be delivered and corded by themselves separate from the rest . . . the said twentieth part shall wholly consist of good free-stone, or of that species of lime-stone which is fit and proper to be cut, chiseled and shaped into the arches, curves and angles of an aqueduct."[21]

The prices to be paid varied between $4.50 and $6.00 per cord, depending on the site to which the material was to be delivered. Contracts in 1818 were priced by the ton and weighed at the place of delivery. Prices varied from 81 cents to $1.31 per ton. The price of masonry measured in place was evaluated by the *perch*. "The prices given for the masonry of the locks, inclusive of stone-cutting, etc., is three dollars per perch of sixteen and a half solid feet, except in one case, where eight dollars per perch, is given for the first foot solid measure, on the inside face of the walls, and one dollar and fifty cents, per perch, for the residue of the lock walls,"[22] The legal definition of a perch was 24¾ cubic feet, but masons generally used the measure of 16½. This custom still prevailed at the end of the 19th century; F. E. Kidder, writing in *Building Construction and Superintendence* in 1896, stated that he had never found a locality where the masons used the legal measure.[23] For the work on King's Chapel, the perch was defined as 1 foot high, 16½ feet long, and 4 feet thick, in a contract dated July 26, 1749. The granite delivered at a Boston wharf was measured by the ton. Cut stone for string courses, moldings and similar trim was usually measured and priced by the linear foot, due allowance being made for the thickness, width and degree of finish.

WORKING AND FINISHING

American stoneworkers, continuing an established European custom, were distinguished according to their tasks: *Quarriers* extracted and roughly shaped the blocks; *rough-masons* "dressed" or finished blocks and cut straight moldings; *freemasons* carved the more intricate shapes and cut curved moldings; *layers* or *setters* placed the blocks in a building. All of these specialists had laborers to assist them. A *master-mason* directed the stonework. He checked it for accuracy, made full-size templates and drew details when these were not furnished by an architect. He often acted as contractor for all of the masonry work and sometimes assumed the contract for a whole building. A stoneworker learned his craft under the direction of a master, to whom

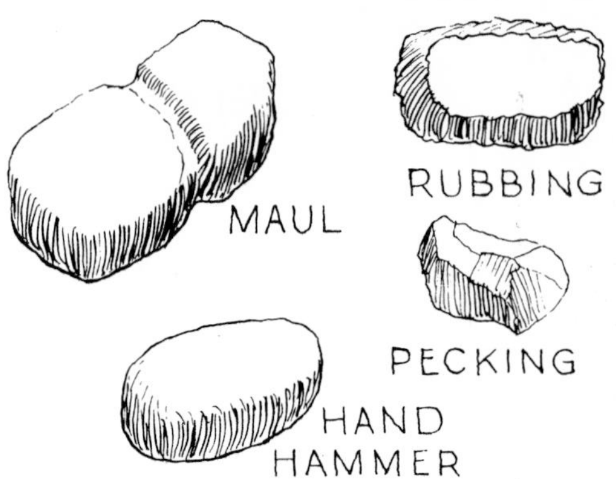

STONE TOOLS. Mauls 5 to 10 inches long; hammers 3 to 5 inches long; pecking stones 2 to 3 inches long. Chapin Mesa Museum, Mesa Verde, Colo.

he was apprenticed for a term of three to seven years.

Long before Europeans settled in North America, builders in the southwestern region of what is now the United States were using stone tools to break and shape stones for masonry walls. At the Chapin Mesa Museum in Mesa Verde, Colo., mauls, hand hammers, pecking stones and rubbing stones that native builders used about 1100 A.D. are on display. Walls built of sandstone during that period often show faces that were broken off evenly. Sandstone breaks cleanly when laid over a straightedge and struck repeatedly with a heavy hammer. Other sandstone walls display faces that were pecked to a handsome finish.

Splitting, Dressing and Squaring. The early colonists in the eastern United States rarely attempted to give a fine finish to stone. In regions where gneiss was found, notably southeastern Pennsylvania, it was split with a stone ax along the rift and broken along the grain. The surfaces obtained in this manner were reasonably even and were approximately perpendicular to each other. The plane of splitting became the *bed* when gneiss was laid in a wall, and the plane of breakage formed the *face*. The third edge of a piece, the *joint*,

Pecked finish achieved by the use of stone implements before 1200. Balcony House, Mesa Verde, Colo.

was usually irregular and oblique. Limestone and sandstone, especially when their strata were prominent, were similarly split and broken. Objectionable protrusions which remained were knocked off with a hammer or an ax. These simple methods sufficed for the finishing of walls, but for sills, thresholds, steps and corners a higher degree of finish was often desirable.

In contrast to the work described previously, which was often done with simple tools by workmen with little or no formal training, the dressing of stone for larger buildings and more finished effects required a greater degree of skill and a number of tools, each adapted to a particular operation and kind of stone. There were five basic methods, all of ancient origin: (1) hewing with an ax or pick, (2) hammering with an ax or hammer, (3) working with a chisel driven by a mallet or hammer, (4) sawing and (5) rubbing with an abrasive. In general, the harder stones, such as granite, were hammered; the softer ones could be hewn and chiseled. Although all hand finishing demanded skill, the use of chisels required more skill than the use of the ax, hammer and pounding stone. Sawing was done to obtain thin slabs of marble, especially for fine veneering and ornamental purposes. Rubbing was usually done only after other methods had been used to approximate the final surface.

Whatever tools were employed, the first step in squaring a block was to *banker* up the widest bed or surface. The workman drew a straight line at one edge and then *pitched* or *smalled* off the debris or waste above it; a *pitching chisel* was the tool used for this purpose. This *first draft* was then refined with a cutting chisel or ax. Next, a draft was made on an adjoining edge of the surface, perpendicular to the first one, and verified with a square. These two drafts defined the plane of the finished surface. By careful use of straightedges and by sighting, the remaining two drafts were cut. The rest of the surface was then reduced to the degree of uniformity desired, with a *point*, hammers, chisels or a combination of tools. The surface just completed became the *bottom bed* of the stone. Then the *face* of the stone was dressed. After that the top bed was finished, then the *end joints* and finally the *inner joint* or surface. The dressing was tested by passing a straightedge smeared with red ocher or chalk over the face of the stone. The color rubbed off onto the high points and indicated to the workman the areas that needed to be further reduced.

Several tools were used in sequence to finish a stone. The work could be terminated at any stage that corresponded to the desired degree of finish. The nature and hardness of the material determined what tools were employed, although there

TABLE 5—WORKING AND FINISHING STONE

Hard Materials (Granite, for example)	Soft Materials (Limestone, for example)
1. Split with drill holes and wedges (at the quarry).	1. Split with drill holes and wedges and/ or "scabble" with a pick (at the quarry).
2a. Establish corners with a pitching chisel and/or	2a. Establish corners with a pitching chisel and/or
2b. "Draft" the margins with ax or chisel.	2b. "Draft" the margins with a tooth chisel.
3. Reduce the surface with a point to within one-half inch of the desired plane.	3. Bring the surface to a rough finish with a tooth chisel.
4. Reduce the surface with an ax or pean-hammer to within one-eighth inch of the desired plane.	4. Cut off the ridges with a wide chisel (called "tool"); the grooves made by the tooth chisel usually show to some extent.
5. Finish with hammers of increasing fineness to the desired degree.	5. If desired, the surface can be rubbed smooth with an abrasive.

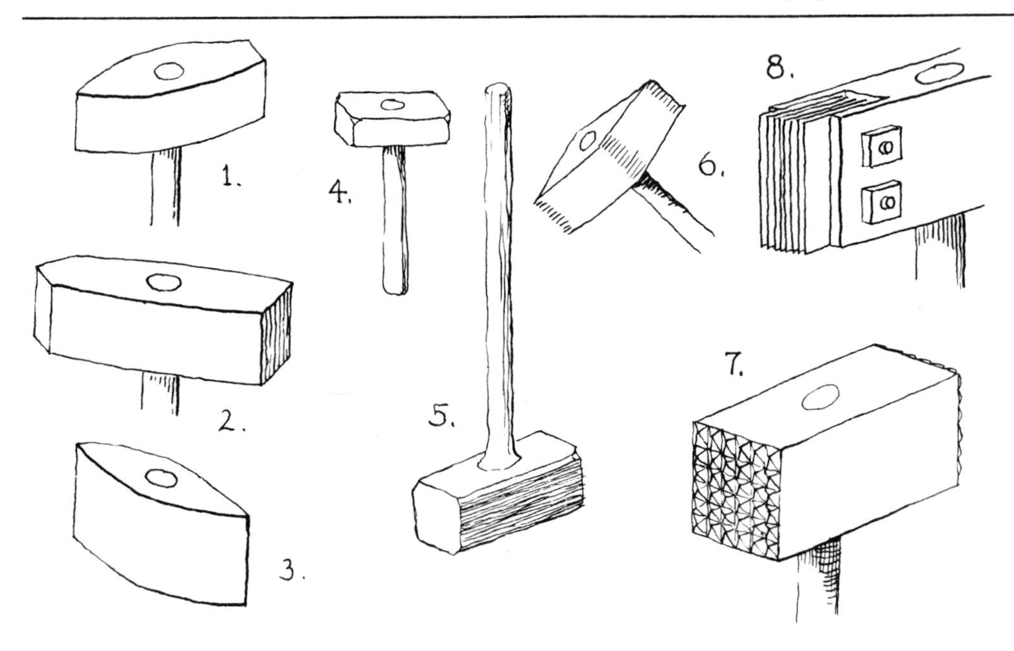

AXES AND HAMMERS.
1. Face hammer.
2. Face hammer; 8 inches long, 2½ inch cutting edge. Mercer Museum, Doylestown, Pa.
3. Ax or pean-hammer; 5¾ inches long, 3 inch cutting edge. Mercer Museum.
4. Hand hammer.
5. Sledgehammer.
6. Ax or pean-hammer.
7. Bush hammer.
8. Patent hammer.

was some freedom of choice. Working and finishing different types of stone might consist of the steps outlined in Table 5.

Each tool had a specific function and method of application. Stone working tools have remained basically the same for many centuries, some even from ancient Roman times. The marks left on the surface of a stone by the last tool used, or the last two tools, can frequently be distinguished if the material is not too weathered. Such marks have always been an important factor in the apparent texture of the material.

Sledgehammers. Sledgehammers weighed from 10 to 25 pounds. They had two square faces and a long handle; they were intended to be held by the workman with both hands and were swung with considerable force. They were primarily employed at the quarry, for driving drills and wedges, knocking off rough projections and breaking stones. Some heavy hammers were made with one square

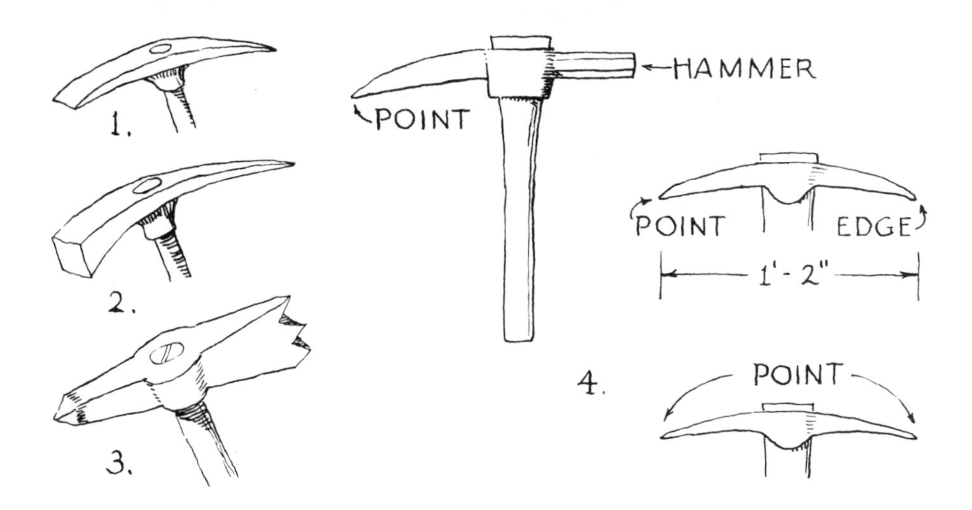

PICKS.
1. Mason's pick with chisel end.
2. Hammer-headed or pole pick.
3. Italian, 16th century.
4. Picks, Mercer Museum, Doylestown, Pa.

face and one cutting face; they were also used at the quarry for rough shaping. The following account written by Chief Justice Shaw in 1859 describes the use of the hammer with a cutting face in New England. ". . . workmen then . . . take the pieces of more regular form and reduce them to smaller and more regular shapes, as wanted for building. This is done by cutting a groove on a straight line with a hammer made with a cutting edge like that of a common axe, then striking it with a very heavy beetle on each side of the groove alternately, until it would crack generally in the line of such groove. This would sometimes split in a line nearly straight, though it would often be irregular."[24]

Drills. Drills or *jumpers* were round, hexagonal or octagonal in section, up to about three feet long, with a chisel-like or flattened point at one end. They were used to cut holes into rock at the quarry, to split off stones and to subdivide large stones. One workman struck repeated blows with a sledgehammer, while another held the drill and rotated it between blows. Hand drills were from 8 to 15 inches long, and were struck with a hand hammer. Sometimes two or three holes were drilled close together and the intervening rock chiseled out. In the late 19th century, steam or compressed air power was adapted to drilling operations.

Wedges. Wedges of hard wood and of copper were used as early as 2500 B.C. but iron or steel wedges have been more commonly used during the last five hundred years. Various tapered shapes have been employed. Rectangular wedges were used in fissures or channels. In round holes a type of wedge called *plug and feathers* was used; a round tapered plug was driven between two hollow semicircular feathers placed in a hole. Plugs and feathers were made in various lengths and diameters. Halbert P. Gillette has described their use in *Rock Excavation*:

With plug holes only 5 ins. deep a block of granite 6 ft. thick can be split. . . . For granite blocks 3 ft. thick, a hole 2½ or 3 ins. deep will suffice. Some limestones also break remarkably well with shallow plug holes, but marbles and sandstones as a rule require deep holes. . . . In most sandstones . . . the holes are usually 1¼ to 2 ins. diam., and . . . of a depth equal to ⅔ the thickness of the stone . . . spaced 4 to 16 ins. apart.

By timing a number of masons at work splitting granite blocks 24 to 30 ins. thick, I found that each man drilled each hole (5/8-in. diam. × 2½ ins. deep) in a trifle less than 5 mins., by striking about 200 blows; and it took about 1 min. for placing and striking each set of plug and feathers. Blocks 30 in. long, with four plug holes, were drilled and split . . . in 24 mins. on the average. At this rate, a good workman can drill and plug 80 holes in 8 hours.[25]

Picks. The mason's pick was an implement similar to an ordinary pickax but shorter and stouter than those used for digging. It was about two inches thick at the eye (through which the handle passed), from fifteen inches to two feet long and pointed on both ends. It was used mostly for rough dressing of sandstone and limestone at the quarry. The block of stone was placed so the face being worked on was not quite vertical, leaning away from the workman, who held the pick handle with both hands and struck the stone with glancing blows, causing pieces to *spall* off. The strokes of a pick ordinarily ran parallel to each other and diagonally to the margins of the block. By examining the finished surface, one can easily distinguish the pits made by the point of a pick from the irregular areas between them.

The pick was often called the *scabbling pick*, and the operation *scabbling*. The finish was called *pick-dressed, picked, scabbled, scappled, nigged* or *nidged*.

The pick assumed various forms. If one end had a square face, the tool was called a *pole pick* or, in heavier form, a *cavil*. Sometimes a cavil weighed 20 pounds or more. Other picks had a blade (like an adze) at one end, or one end might be toothed.

Points. Points were round or octagonal in section, about 12 inches long when new and sharpened to a pyramidal point. They were used on hard stones to remove material quickly. A point was held in one hand and struck with a hammer held in the workman's other hand; it had to be sharpened frequently and was discarded when worn down to a length of about five inches. The hammers employed weighed from two to five pounds.

When used on granite, the point removed material down to one-half inch from the finished plane. The remainder of the dressing was done by hammering. Points and narrow chisels (about one-quarter of an inch wide) were used to finish sandstone and, less frequently, limestone. When the grooves were con-

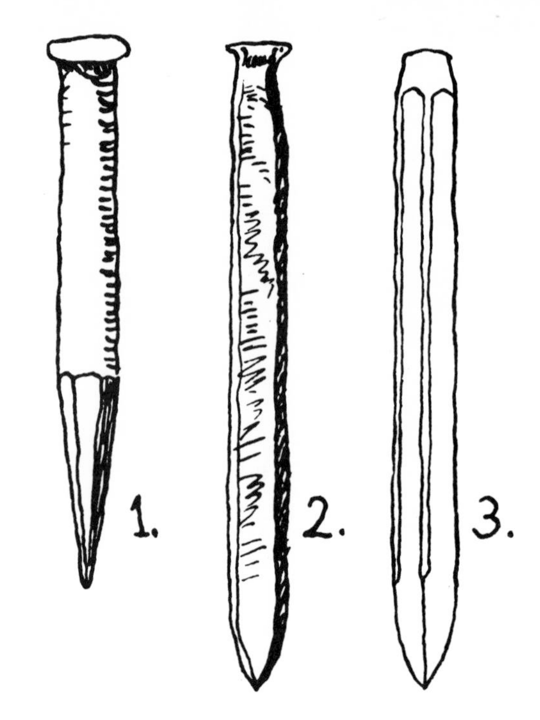

POINTS.
1. Type used since ancient Roman times.
2. Italian, 16th century.
3. American, 19th century.

Detail of an 1828 sandstone bridge. The upper portion was dressed with a pick (scabbled), while the string course (at bottom) was chiseled. Blaine, Ohio.

tinuous (and parallel) the finish was called *broached work.* The grooves usually extended in a diagonal direction but could be horizontal or vertical. The operation was called *broaching.* When the grooves were not continuous and were one inch apart the surface was said to be *rough pointed;* when the grooves were half an inch apart, it was *fine pointed.* The grooves were from one-quarter to one-half inch deep.

Pitching Chisels or Pitchers. The pitching chisel was a rather heavy chisel, about one and one-eighth inches in diameter, with a cutting edge that spread to two and one-half inches. The cutting edge was not sharpened. It was flat, about one-half inch thick. The only use for the pitching chisel was to *pitch off* the edges of a stone that was being squared up. If the face of the stone was not given any additional finish it was said to be *rock faced.*

Axes or Pean-Hammers. The mason's ax or stone-ax, often called a *pean-hammer,* was about ten inches long; it had cutting edges at both ends, which were about four inches long. The handle of the ax was up to 30 inches long. If the cutting edges were divided into teeth, the implement was called a *tooth ax.* The workman brought the cutting edge of the ax down directly onto the stone with a hammering motion; for this reason the work was called *hammering.* Crystals directly under the cutting edge, which struck the

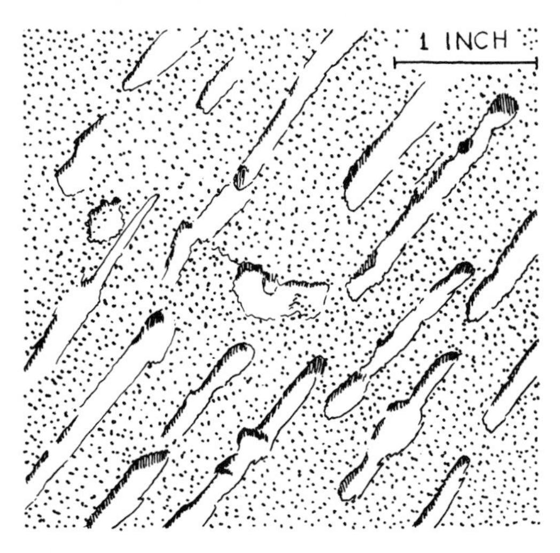

AX MARKS. Hammered finish on limestone, 12th century, Triforium gallery of north transept, Winchester Cathedral, England.

Point work on sandstone. Margins are chiseled. St. Mark's Episcopal Cathedral (1870), Salt Lake City, Utah. Richard Upjohn, architect.

material with blows perpendicular to the surface, were crushed by hammering. The disturbed crystals that remained attached to the stone became more susceptible to the forces of weathering than those that were not struck directly.

Hammering probably originated with ancient Egyptian use of pounding-stones. In the Middle Ages, iron axes were customarily used for dressing stones by hammering and may have been used in other ways to finish stone surfaces. Traditional methods of using the ax or pean-hammer were brought to America by workmen familiar with the medieval buildings of England. L.F. Salzman, describing medieval practice in *Buildings in England Down to 1540*, mentions the ax:

> . . . the most momentous change in the working of stone was the introduction of the use of the chisel not only for carving, but, to a rapidly increasing extent, for dressing stones, which occurred in the second half of the twelfth century. . . . In the matter of surface dressing, although documentary evidence is absent, we have evidence of the stones themselves, with their unmistakable axe-tooling, diagonal on flat surfaces and vertical on the rounded surface of pillars. . . . Nor is there any implement named in building accounts which can be identified with the drag, used to give a fine finish to the surface. My own impression, based partly on illustrations, is that the drag of the Middle Ages was simply the blade of the mason's hammer-axe, used with a drawing motion. . . . There were two main types of axes in use: the mason's axe proper, as still used, had two vertical edges; the hammer-axe was like a small mattock, having one horizontal edge and on the other face a hammer-head.[26]

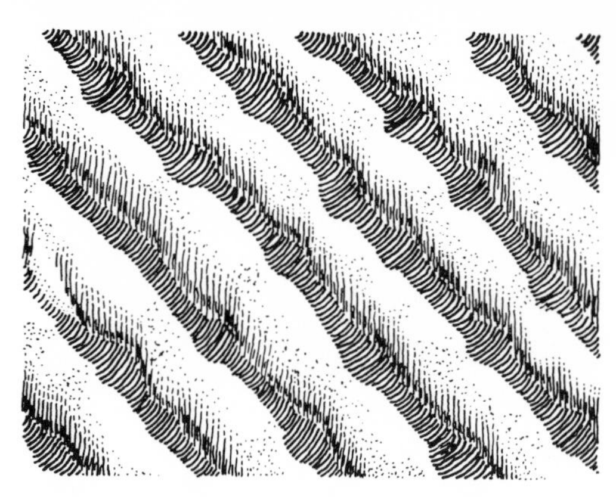

BROACHED FINISH. Marble, worked with a point. Sometimes the cuts are made horizontally. St. James Episcopal Church (1831), Arlington, Vt. (Drawing is one-half actual size.)

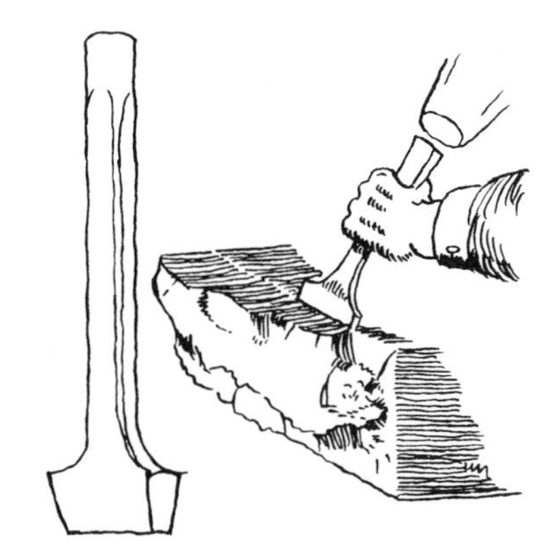

PITCHING CHISELS. American, 19th century. Pitching chisels have a flat face instead of a cutting edge.

There is evidence that the pean-hammer was used in the United States from the mid-18th century, especially in finishing granite. The characteristic grooves of the pean-hammer can be seen on the surface of stone, but they are often difficult to distinguish from those made by a *patent hammer*.

Bush Hammers and Patent Hammers. The bush hammer was used in Europe for several centuries before it was brought to America. It had two square faces. Each was divided into a number of regular pyramidal points, spaced from about two to an inch to six to an inch. With each blow of the bush hammer, the workman made several rows of regular dents in a stone. For the finest work, care was taken to make each stroke in the same direction; at other times, the direction of the marks is seen to vary, often following an arc with a radius the length of the stone-cutter's arm. The crushing of grains in the stone, under each point of impact, can be noted on close inspection. It is particularly visible in limestone, showing up as small spots perhaps one-eighth of an inch in diameter and lighter in color than the rest of the surface.

The bush hammer was widely used in the United States by the early 19th century. The author has noted the early 19th-century use of the bush hammer on sandstone and its popularity for finishing limestone by the middle of the century. A bush-hammered finish has an attractive stippled appearance which contrasts nicely with the nearly invariable chiseled margins of each stone.

PATENT-HAMMER FINISHES.

Left: Sandstone, 8 cuts to the inch. Alexander Brown House (1895), Syracuse, N. Y.

Right: Granite, 12 cuts to the inch. John Crouse Mausoleum (1884), Oakwood Cemetery, Syracuse, N. Y.

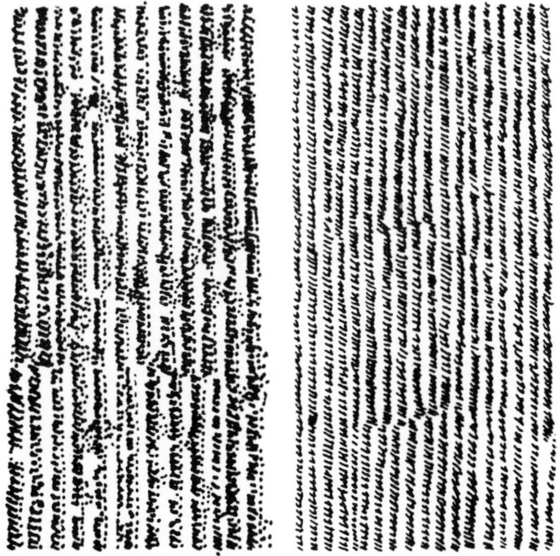

Rock-faced surface of Onodaga limestone. A pitching chisel was used in finishing. Horizontal fissures are common in this type of limestone.

The patent hammer, invented by Joseph Richards, was introduced about 1831.[27] Documentary references to it can be misinterpreted because it was also called a bush hammer, especially in the eastern states, and an ax hammer. By the latter part of the 19th century it became the usual tool for hammering granite and was also used to finish sandstone. The cutting face of a patent hammer consisted of several thin blades clamped together into an assembly about one inch thick. These blades could be removed for sharpening. Depending on the thickness of each blade, there were between five and twelve blades to the inch, each about two and three-quarters inches long. Thus, a number of parallel cuts could be made at the same time in a stone. It was customary to begin with coarser cuts, then change successively to finer cuts at right angles to the previous ones. It was unusual to specify finer than eight-cut work on buildings but twelve-cut finishes were applied to expensive funerary monuments. On wall faces, the cuts were vertical but on curved and molded surfaces they were parallel to the length of the molding (i.e., following the elements of the cylindrical surface).

Crandalls. Another kind of finishing hammer called a *crandall* came to be used in the late 19th century, principally on sandstone. The crandall had a somewhat flexible iron handle about two feet long. A number of steel rods sharpened to a point or to a chisel edge were inserted in a slot three inches long and held in place by a steel wedge or key. These rods could be removed for sharpening.

Bush-hammered sandstone water table with chiseled margins. The wash was also chiseled. Wasatch Stake Tabernacle (1887), Herber City, Utah.

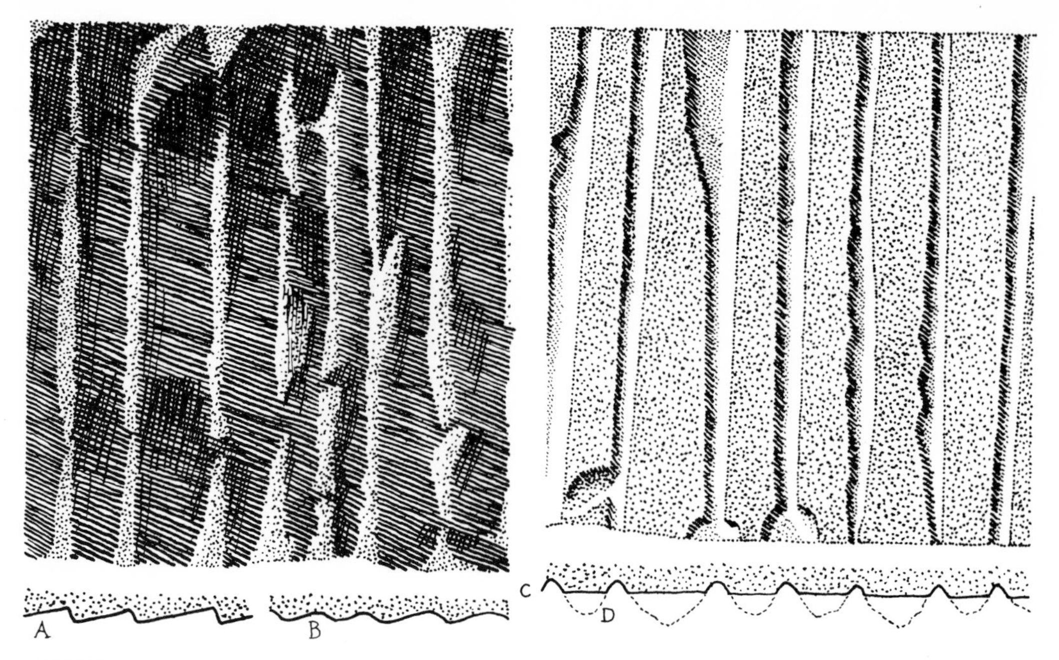

CHISEL WORK.
Left: Sandstone string course, bridge on the National Road (1828), Blaine, Ohio. When freshly dressed, a section of the surface probably resembled figure A; after weathering for 143 years it resembles figure B.
Right: Sandstone window sill, Friends Meeting House (1814), Mount Pleasant, Ohio. The stone was brought to a rough surface with a point and finished with a wide chisel which removed the ridges indicated at D but left the bottom of the grooves at C intact.
(Drawings are one-half actual size.)

An apprentice at Chichester Cathedral Work Shops in England dresses soft limestone with a chisel and wooden mallet. Note the rasp and straight edge on the bench. In this type of restoration work, stones are sawn oversize at the mill and finished by hand in the workshop.

Chisels. The origin of the chisel, as well as the ax, is found in the utilization of pieces of flint, chert and nephrite as cutting edges for implements of wood and bone. Early Egyptian stoneworkers had chisels of stone, copper and bronze, but the potential importance of the chisel as a tool for working stone was not realized until iron and steel became available. Chisels of various widths, with straight cutting edges or with teeth, varied little in their essential forms from ancient times to the 19th century. Several forms of the chisel were used by sculptors and stonecarvers as well as rough-masons. However, a number of small kinds that were available to the decorative worker were not used for plain surfaces.

The chisel was held in one hand, at an angle of about 35 degrees to the plane of the stone. If a chisel was held at a steeper angle, crystals of the stone would be crushed. The chisel was tapped with a wooden mallet or iron hammer, whose weight varied with the softness or hardness of the stone. It was normally used on softer stones. The term *freestone* was used to designate material that could be cut in any direction by a chisel. In finishing stones, the chisel was driven along its path by a series of taps with the mallet or hammer. In work of better quality, a chisel was driven the entire length of its path before being lifted. Workmen with chisels received a higher rate of pay than those with hammers.

Stones were usually cut before the groundwater in them, called *quarry sap*, had dried out. In this condition they were softer and easier to work with a chisel or other tool.

Tooth chisels with serrated cutting edges were used for the preliminary reduction of soft stones. The final surface was cut with wide chisels having straight edges; these removed the tops of the ridges left by tooth chisels. The grooves remained as shallow parallel marks running vertically along the surface of the stone although if the stone was very soft these marks could be effaced. The grooves at the margins of each stone were made perpendicular to the edges; these margins were from one to two and one-half inches wide. The straight chisels used for the final operation varied in width: One between two and three inches wide was called a *drove*. One from three and one-half to four and one-half inches wide was called a *tool*. The surfaces they produced were called *drove work* and *tooled work*, respectively.

One type of chiseled surface frequently seen on buildings constructed during the first half of the 19th century in the United States consists of a series of narrow vertical bands, which on first sight resemble the fluting of a column. On closer inspection, one may see that the profile of the stone more nearly resembles the teeth

CHISELS.
1. 7 inches long, 1 inch octagonal, 1½ inch cutting edge, Mercer Museum, Doylestown, Pa.
2. Tooth chisel; 5 inches long, ¾ inch octagonal, 1¾ inch cutting edge with 7 teeth. Mercer Museum.
3. Splitting chisel. Chiefly used for soft stratified stones; occasionally used for carving granite.
4. Narrow chisel, 19th century.
5. Tooth chisel; Italian, 16th century.
6. Tooth chisel, 19th century.
7. Tool; 3½ to 4½ inches wide. Also called boaster *or* bolster.
8. Drove; 2 to 3 inches wide.

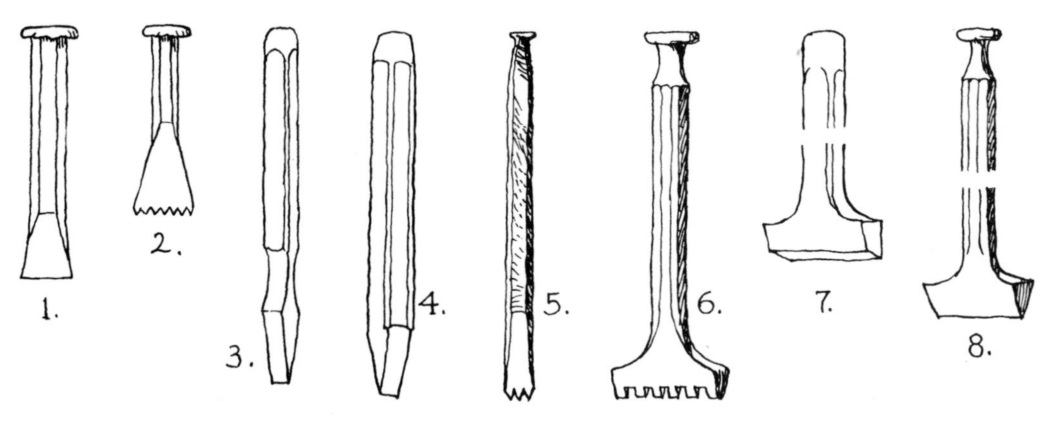

of a rip saw. Bands from three-eighths to five-eighths of an inch wide, at an acute angle to the plane of the wall, are separated by narrow bands from one-eighth to three-sixteenths of an inch wide, at an obtuse angle to the plane of the wall. When weathered, these surfaces assume a wavy form. They appear to have been dressed with a straight chisel perhaps three-quarters of an inch wide. This type of dressing is seen on the risers of stone steps, on the face of copings and sills and on string courses, as well as on entire wall surfaces.

Abrasives. After the surface of a stone was worked to a plane with chisels or hammers, it was brought to a perfectly smooth finish by rubbing with an abrasive block of hard sandstone or with any other kind of block and wet sand. Mills and factories rubbed and polished stones by means of water or steam power from the early 19th century on. Heavy cast iron discs were rotated on top of the stone being finished while finer and finer abrasives were fed under the discs.

Saws. The use of abrasive grains in sawing and drilling stone probably had its origin in the making of jewelry. Ancient Egyptian beadmakers used a bow-drill, reeds and an abrasive. Pre-Hellenic Greeks used a toothless copper blade and grains of emery to saw hard stone. A copper saw set with teeth of emery, found in the ruins of the Palace at Tiryns, was used for cutting limestone.[28] Sawing has been practiced chiefly to obtain the greatest number of usable pieces from each block of rare, unusually beautiful or expensive stone, such as marble. Thin slabs suitable for veneering and paving were produced with little waste of material.

Handsaws were made from blades of smooth soft iron; a piece of wood attached to the back of the blade served as a handle for a small saw. *Frame saws* were more commonly used. The frame saw resembled its counterpart for cutting wood, except for the absence of teeth. Sand was the most common abrasive material; it was placed in a groove in the stone, where it was moved back and forth by the saw blade. The sand had to be replenished and wet at frequent intervals.

The principle of the frame saw was adapted to industrial use in the marble mills of New England. Several blades or

Cobblestone facing. Stones were gathered from nearby fields. Both horizontal and vertical joints form **V***-ridges. Cobblestone buildings are numerous in western New York but are rare elsewhere. Universalist Church (1834), Childs, N.Y. (Jack E. Boucher for Historic American Buildings Survey).*

Limestone blocks on the front wall were dressed; stones on the end wall have broken faces. Sandstone quoins accent the corner. Belle Grove (1794), Middletown, Va. (William Edmund Barrett)

"gangs" parallel to each other were stretched in a wide frame so that several slabs were sawn at the same time. Weighted pulleys applied the proper amount of pressure on the blades, which were mounted in a horizontal position and moved back and forth by waterpower. Before mills were developed in Vermont, it was the usual practice to select marble from which *sheets* could be split off; these were then worked with chisels.

At first, the production of mill-sawn marble for building purposes was auxiliary to the making of gravestones, for which there was a greater demand. However, once established, the industry became more specialized and building stone was produced separately. Early mills were located where waterpower was available. Some marble quarries in New England were adjacent to sources of power, and their operators were able to combine the processes of quarrying and milling into a single, efficient industry. About 1783, William Shepherd erected a mill for sawing and polishing marble in the vicinity of Philadelphia.[29] In 1804, Eben W. Judd began to operate a mill in Middlebury, Vt.; by 1806 it was producing on an extensive scale.[30] The Middlebury Marble Manufacturing Company, incorporated in 1809, sawed 20,000 square feet of marble slabs during that year and the following one. Other important centers of sawing and polishing marble were New York City and Baltimore. Sills, threshholds, hearths, mantelpieces and other kinds of trim were manufactured in considerable quantities and shipped to dealers in many cities. About the middle of the 19th century, marble articles imported from Europe, especially from Italy, offered serious competition to domestic manufacturers. However, one cannot always be certain whether the "Italian marble" referred to in dealers' advertisements actually came from Italy, or whether it was domestic material resembling Italian varieties. Other competition came from the manufacturers of *marbelized slate* articles, who gave their product a baked and lacquered treatment that closely simulated marble.

STONE WALLS

The subject of masonry work in buildings is so vast that only a brief summary can be given here. Joints in masonry walls will be considered primarily in the chapter on mortar.

Hoisting Stones. Stones too heavy to be lifted into position by hand were hoisted by methods similar to those described above in quarrying. Some were pulled by hand up ramps supported by scaffolding; others were so large and heavy that they required horse-drawn carts. The *lewis*, a

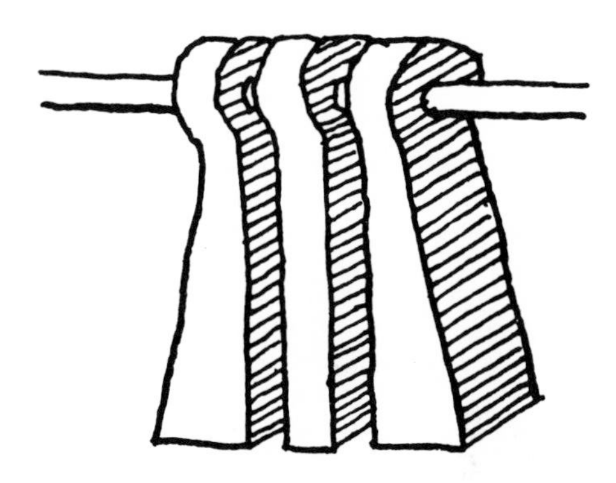

LEWIS. This device used for hoisting stones has been known since Hellenic times. It is an iron tenon, made in sections, which can be fitted into a dovetail mortice.

STONEWORKING.
1. Finishing with a chisel and wooden mallet.
2. Dressing with a pick. This is called scabbling or scappling.
3. Sawing marble into slabs.

device used since 500 B.C. in Greece, was commonly employed to attach the hoisting ropes to a piece of cut stone. The lewis consists of an iron tenon, made in sections, which can be fitted into a dovetail mortise in the top of the stone and easily removed section by section when the stone is in place.

Bonding. Garden walls and the foundations of simple farm buildings often had *dry walls*, or walls laid without mortar. Most stone buildings, however, depended on mortar. In either case, attention had to be given to selecting and laying stones so as to *bond* or interweave them, in order to make the wall strong. Bonding is most apparent on the exterior face of the wall but the stones on the interior of a wall must also be bonded. There are various ways or patterns of bonding, for which terminology differs with time and place. The author has used the terms as they are defined by A.L. Osborne in *A Dictionary of English Domestic Architecture*.

Lumps of stone for building work are used either uncut and irregular as they come from the quarry, roughly cut to rectangular faces, more carefully shaped and selected to lie in horizontal courses, or cut and shaped so that the edges of the blocks form accurate rectangles, the visible faces being rubbed true and smooth. The last type is known as *ashlar;* the others are all forms of *rubble masonry*. . . . Rubble masonry may be divided into two main kinds, in which the blocks are either uncut or roughly squared. Of the former, *random rubble* consists of blocks of various shapes and sizes laid with thick mortar joints . . . while *coursed random rubble* refers to the use of uncut blocks selected to bed horizontally, and is found in districts where the stone splits or "cleaves" regularly. . . . examples remain in which the stones are cut almost as carefully as in ashlar work. . . *cut rubble* is a term sometimes used for these types. [31]

Thick masonry walls were often finished with a non-load-bearing surface *veneer* of finer or more carefully worked material. For example, rubble and brick walls were often veneered with ashlar.

Laying Tools. In laying walls, a mason used an iron or steel *trowel* shaped somewhat like a diamond (or rhombus). The handle was attached to a Z-shaped iron rod at the heel of the implement; the point was rounded. The trowel was used primarily to apply mortar and also to tap small stones into position. *Levels* were made of wood, in the form of an A or an inverted T: a plumb bob was suspended from the top. When the plumb line hung in alignment with a mark at the center of the bottom of the apparatus, the two feet were level. Verticality was checked by an instrument operating on the same principle. Levels of this type are of ancient origin but were used well into the 19th century. A wooden or iron *square* was used to verify right angles. A *compass* with steel points was employed to draw arcs of small radius; steel points were used to draw straight *scribe lines* on stone. *Lines* were cords stretched between stakes to mark important corners.

Sandstone coursed rubble with a pecked finish. Sun Temple (before 1200), Mesa Verde, Colo.

Leveling lines were stretched between corners and raised as the walling progressed. *Snap lines* were cords to which chalk or ocher was applied so that a reference line could be transferred to a wall or other surface.

DETERIORATION OF STONE AND STONE WALLS

Causes. Structural failures in masonry construction are manifested by cracking, uneven settlement, bulging, deterioration of mortar and other visible signs. Such failures frequently need to be remedied. Another problem is the deterioration of the stone itself. In regions where the atmosphere is polluted, this condition is becoming increasingly widespread.

Every kind of stone is more or less porous; it absorbs moisture from a damp atmosphere, from rain, from groundwater and from condensation on the interior of the building. If there are soluble salts within a stone, or if some are introduced by *rising damp* (moisture carried upward through a wall by capillary action) from the ground, they may be carried toward the face of the wall. If these salts crystallize within the pores of the stone, the action may cause the surface to break off; if they are carried to the surface and then crystallize on it, unsightly efflorescence is formed.

All rainwater contains some dissolved carbon dioxide; that in congested metropolitan and industrial areas also contains sulphuric and nitrous oxides, making it especially damaging to stone. Dirty surfaces attract moisture thus making them particularly vulnerable to disintegration or "stone diseases." If water in the pores of a stone freezes, it can cause portions of the surface to break away. Similarly, water that penetrates joints and cracks can cause serious damage by by freezing. Most kinds of stone are damaged by exposure to the intense heat of a fire. Marble and limestone are calcined, and the surface of granite breaks off. Even when heat damage is not apparent to the eye it may cause incipient cracks that later become larger.

Remedial Measures. During the repair or restoration of a building, seriously damaged stones are usually replaced. The introduction of *dampproofing courses* can protect a wall against rising damp, but moisture from other sources cannot be completely controlled. Some types of chemical treatment have been found effective in protecting valuable works of stone sculpture in sheltered locations but the high cost of such treatments restricts their use on the exterior of buildings. Other chemical treatments may appear to be effective for a short time but may gradually weaken and damage stone surfaces.

Rock-faced sandstone, random ashlar. Drill marks are visible. Starrucca Viaduct (1846), Lanesboro, Pa.

Monolithic granite columns, 20 feet, 9 inches high. These were brought to Boston by canal from a quarry in Chelmsford, Mass. Quincy Market (1825), Boston, Mass. Alexander Parris, architect.

Gneiss rubble with granite quoins. The granite was dressed with a point. Swarthmore, Pa.

Restored 18th-century sandstone and granite rubble. Lafayette's Headquarters, Chadd's Ford, Pa.

Coursed rubble of roughly squared brownstone. St. Peter's Church (1767-69), Middletown, Pa.

Shaly sandstone rubble with alternating courses of small and large stones. Chaco Indian Pueblo (c. 1100), Aztec National Monument, N.M.

Coursed sandstone rubble. Note the long thin pieces of stone. Leiby Farm (c. 1830), Virginville, Pa.

Monolithic granite piers and lintels in a stone skeleton facade. East India Marine Hall (1824), Salem, Mass.

Aquia Creek sandstone trim, rubbed finish. Aquia Church (c. 1750), Stafford, Va.

Seneca Creek sandstone ashlar, pecked and fine-point finishes. Smithsonian Institution (1847), Washington, D.C. James Renwick, architect. (Renwick commissioned accelerated weathering tests on 19 types of stone before selecting Seneca Creek sandstone for the building. The different types of stone were alternately boiled in sulphate of soda and dried. The Seneca Creek stone has endured well in the building.)

Rock-faced sandstone ashlar with an angle-draft at the corner. Grace Episcopal Church (c. 1850), Honesdale, Pa.

BANKERING UP A BLOCK OF STONE.

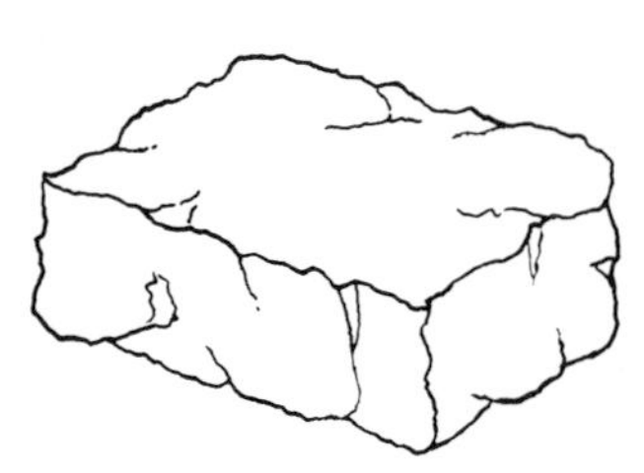

1. Roughly squared stone as delivered from quarry.

2. The first edge is pitched off.

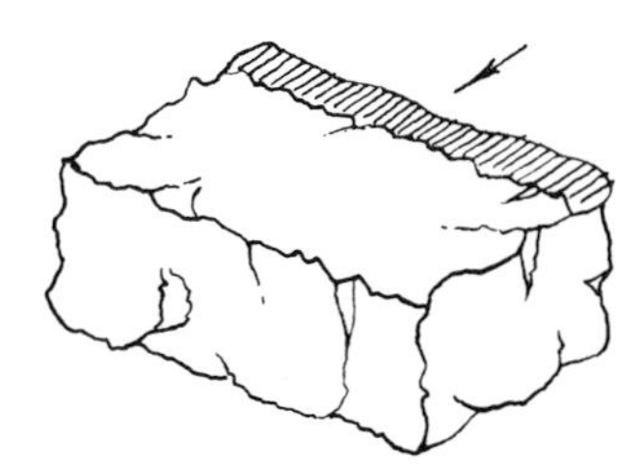

3. The first draft is cut.

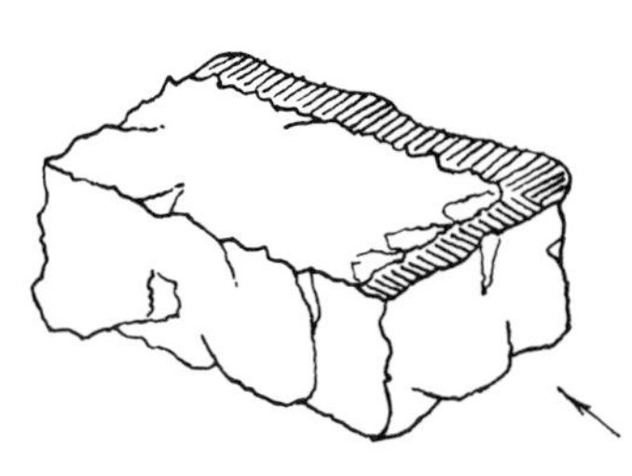

4. The second edge is pitched off and drafted.

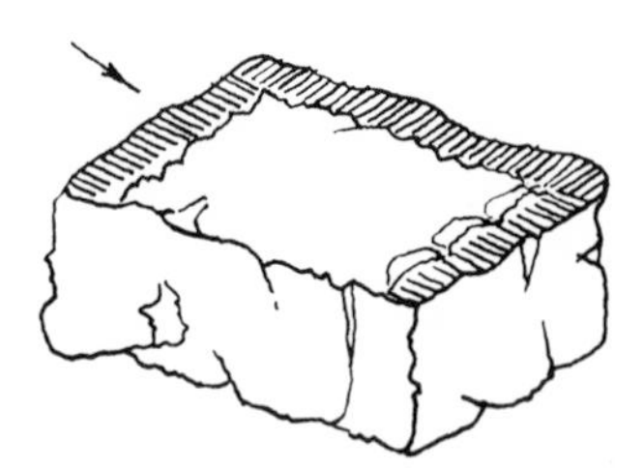

5. The third edge is pitched off and drafted.

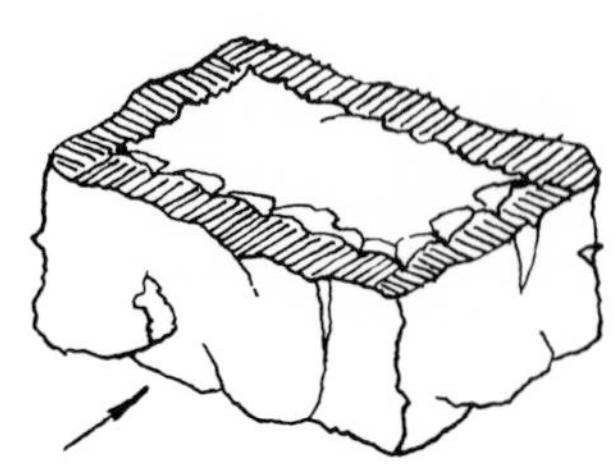

6. The fourth edge is pitched off and drafted.

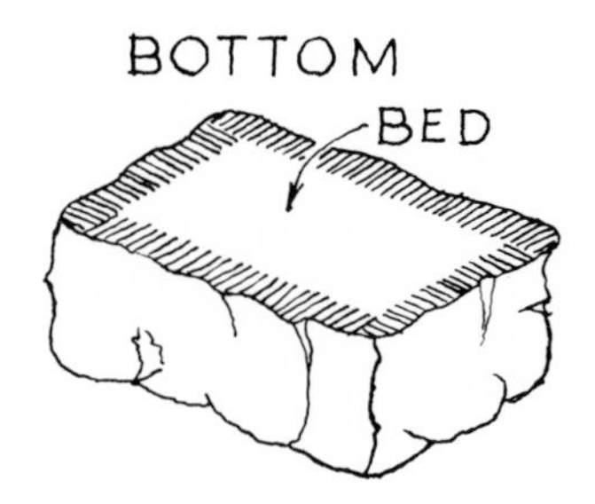

7. The first surface is dressed.

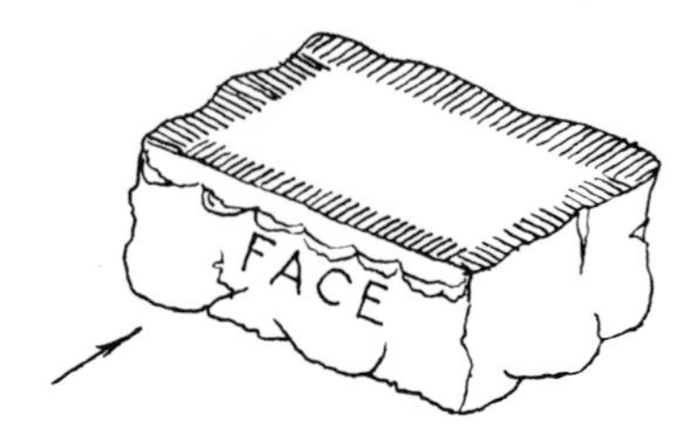

8. The first edge of the face is pitched off.

9. The block of stone is turned.

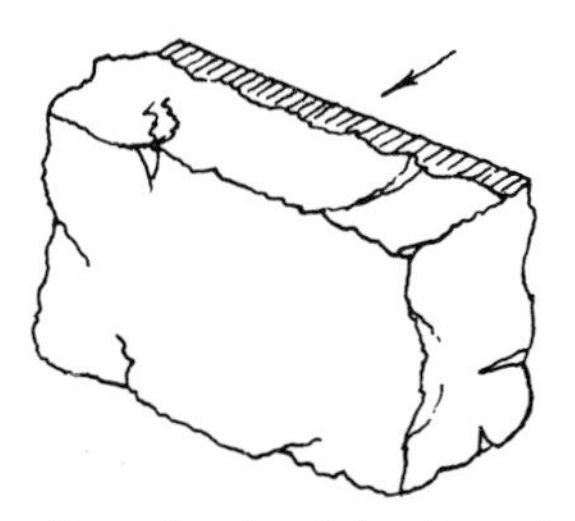

10. The first draft of the face is cut.

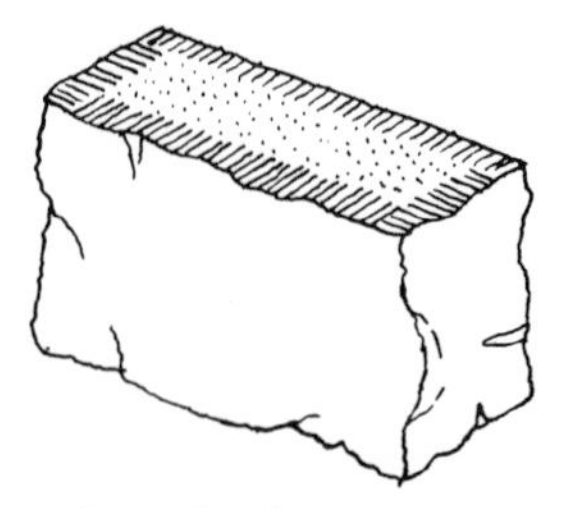

11. The other drafts are cut and the surface is dressed.

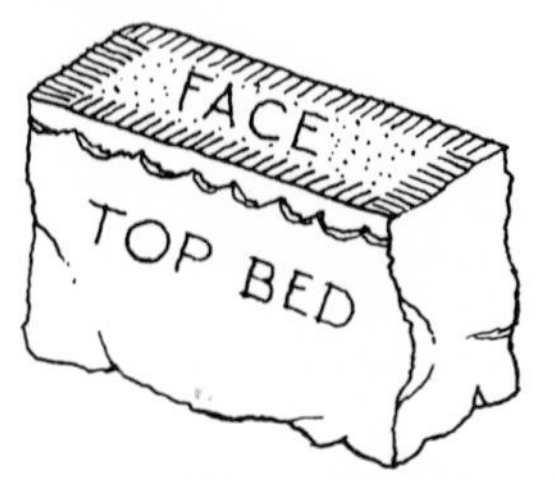

12. The first edge of the top bed is pitched off.

13. *The block of stone is turned.*

15. *The other drafts are cut and the surface is dressed.*

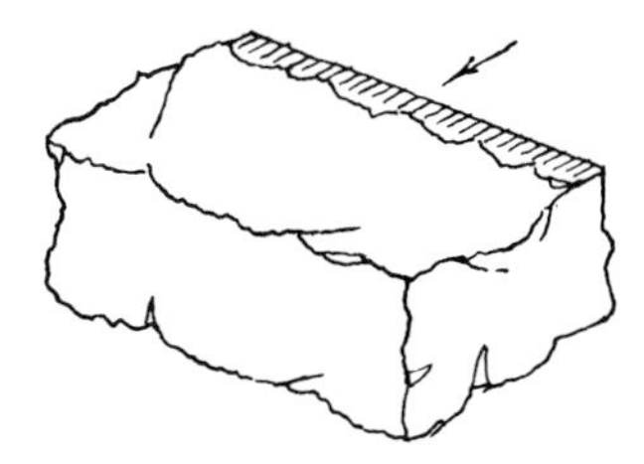

14. *The first draft of the top bed is cut.*

16. *Similarly, the two joints and the back are dressed in succession.*

FOOTNOTES FOR CHAPTER I

1. Edward B. Mathews, "An Account of the Character and Distribution of Maryland Building Stones together with a History of the Quarrying Industry," *Maryland Geological Survey,* vol. 2 (Baltimore: Johns Hopkins Press 1898), p. 152.
2. *Public Documents Relating to the New York Canals* (New York: State of New York, 1821), p. 216.
3. Charles E. Foote, "Bluestone," in *The History of Ulster County, New York,* ed., Alphonso T. Clearwater (Kingston, N.Y.: W. J. Van Deusen, 1907), p. 542.
4. Oliver Bowles, *The Stone Industries* (New York: McGraw-Hill, 1934), pp. 148–49.
5. Norman M. Isham and Albert F. Brown, *Early Connecticut Houses,* (1900; reprint ed., New York: Dover, 1965), pp. 173–74.
6. Ralph W. Stone, "Building Stones of Pennsylvania," *Pennsylvania Geological Survey,* 4th ser., bulletin M 15 (Harrisburg, Pa.: Topographic and Geologic Survey, 1932), p. 97.
7. Ibid., p. 70. Quarry observed by author in 1970.
8. Isham and Brown, op. cit., p. 176, quoting David D. Field, *Statistical Account of the County of Middlesex in Connecticut* (1819).
9. Carl O. Dunbar, *Historical Geology,* 2nd ed. (New York: Wiley, 1961), passim.
10. Compiled by the author from many sources.
11. Albert C. Manucy, *The Building of Castillo de San Marcos,* National Park Service Interpretive Series, History no. 1 (Washington, D. C.: 1942), p. 9.
12. J. A. Bownocker, "Building Stones of Ohio," *Geological Survey of Ohio,* 4th ser., bulletin 18 (Columbus, Ohio: 1915), pp. 108 ff.
13. Vitruvius, *The Ten Books on Architecture,* trans. Morris Hicky Morgan (1914; reprint ed., New York: Dover, 1960), bk. 2, chap. 7, p. 5.
14. George P. Merrill, *Stones for Building and Decoration,* 3rd ed. (New York: Wiley, 1910), p. 393, footnote, quoting Darlington's Memorandum of Bartram and Marshall.
15. Albert D. Hager, "Economical Geology of Vermont," *Report on the Geology of Vermont* (Claremont, N.H.: 1861), p. 757.
16. Adapted from Arthur W. Brayley, *History of the Granite Industry of New England,* vol. 1, chap. 2 (Boston: National Association of Granite Industries of the United States, 1913).
17. Ibid. Brayley indicated that the descriptions of machinery were quoted from Solomon Willard.
18. Ibid.
19. Ibid., pp. 68–69.
20. F. W. P. Greenwood, *A History of King's Chapel in Boston* (Boston: Carter, Hendee & Co. and Allen & Ticknor, 1833), p. 125, footnote. Henry Wilder Foote, *Annals of King's Chapel,* vol. 2 (Boston: Little, Brown, and Company, 1896), pp. 90–91.
21. *Public Documents Relating to the New York Canals* (New York: State of New York, 1821), pp. 300–301.
22. Ibid., p. 342.
23. F. E. Kidder, *Building Construction and Superintendence,* 9th ed., rev. by Thomas Nolan (New York: Comstock, 1910), p. 307.
24. Braley, op. cit., p. 16.
25. Halbert P. Gillette, *Rock Excavation* (New York: Myron C. Clark, 1907), pp. 187 ff.
26. L. F. Salzman, *Building in England Down to 1540* (Oxford: Clarendon Press, 1952), p. 333.
27. Braley, op. cit., p. 87.
28. Charles Singer et al., eds., *A History of Tech-* 1954, 1957), vol. I, pp. 139–40, pp. 189–92; vol. II, p. 29.
29. J. Thomas Scharf and Thompson Westcott, *History of Philadelphia,* vol. 3 (Philadelphia: L. H. Everts & Co., 1884), p. 2229.
30. Samuel Swift, *History of the Town of Middlebury* (Middlebury, Vt.: A. H. Copeland, 1859), p. 335, quoting Professor Frederic Hall, *Statistical Account of the Town of Middlebury* (1821).
31. A. L. Osborne, *A Dictionary of English Domestic Architecture* (New York: Philosophical Library, 1956), p. 60. This excellent dictionary has been out of print for some years.

II. BRICK

INTRODUCTION

History of Brick. Small baked clay units laid in mortar have been used to build walls in various parts of the world since the early days of civilization. "Burned" (baked) bricks have been used as building material for 9,000 years. They have always been manufactured locally wherever suitable ingredients and fuel were available. Where fuel was not in ample supply, sun-dried bricks were made.

The brick buildings of Persia, northern Italy, Flanders, Holland and the Baltic area of Germany are particularly notable. During the Middle Ages, workmen migrated to England from Flanders and, to a lesser extent, from other northern European brickmaking centers. Their techniques of manufacture and their building practices largely formed the brickworking tradition brought by builders to the British colonies in America.

Use of Brick in Early America. By the 16th century, brick was a fashionable building material in England, a country well supplied with stone. The status enjoyed by owners of brick houses was a factor in the use of brick for the construction of the finer houses built by English settlers in America. The Dutch who founded New Amsterdam and other cities in the Hudson Valley were accustomed to brick buildings of excellent craftsmanship in Holland and wanted to duplicate them in their new settlements. In the colonies as a whole, however, the number of brick buildings was small compared to those built of wood, but many of the latter had brick chimneys and foundations, so there was some demand nearly everywhere for the material.

BRICKMAKING

Local Production. Bricks have always been manufactured locally, partly because of their bulk and weight and partly because the clay and sand from which they are made are found almost everywhere. Brickmaking is widely distributed throughout the United States even today. Bricks can be manufactured in improvised temporary facilities and were often made at the building site. Marcus Whiffen mentions this in *The Public Buildings of Williamsburg:*

> The bricks [for the first building of the College of William and Mary] were made on the spot by Daniel Parke, a member of the Council, for 14S. a thousand. . . . The kiln in which the bricks were burnt was found by excavation in 1929. . . .
> There were two bricklayers among the first settlers at Jamestown; six more, and four brickmakers, were included among the tradesmen accompanying Sir Thomas Gates in 1610, and bricks continued to be made in Virginia throughout the colonial period. In the case of a brick house of any size, the bricks might be made for the job on the site or near it. . . .
> Remains or traces of brick kilns have been found in six or seven places in Williamsburg.[1]

When brickmaking became an industry, most cities had at least one brick plant. Bricks made at a local plant could be sold at a lower price than those brought from more distant factories. In some regions, brickmaking plants flourished more than in others, and the products of these plants were more widely shipped to consumers. (In the 20th century, differential freight rates facilitated the shipment of brick and made brick shipped from a distance competitive in price with locally made products.)

Clay. Clay is a material of varying composition formed by the erosion of rocks on the outer crust of the earth. Small particles of clay are carried by streams to river deltas or places along streams where the velocity of the water is low. The clay in any given deposit may have come from various sources. Types of clay found in different localities may differ considerably in their properties; thus the bricks made of material from different clay pits differ accordingly.

The most common kinds of clay minerals are *kaolinite, montmorillonite, illite* and *chlorite.*[2] The molecules of these clay minerals have a layered structure that permits them to absorb varying amounts of water. Montmorillonite is especially noted for its absorbency.

For reasons of economy, clay is used in the form in which it is found; deposits often contain other substances, such as oxides of iron, lime, magnesia and alumina. The presence of these "impurities" is significant. Five to six percent of ferric oxide in the clay produces a characteristic red color; hard red bricks can be produced at a moderate temperature (2,000 degrees F.). The presence of lime produces yellow or greenish-yellow bricks and demands extremely accurate control of the firing temperature. The presence of magnesia and alumina produces buff-colored bricks. Ferric oxide produces various shades between pink and reddish black; ferrous oxide produces shades ranging from green to black. Manganese gives clay a brown color.[3] The brickmaker of colonial times did not understand the chemical composition of clays; he had to rely on observation and practical experience. It is little wonder that the appearance and other characteristics of early bricks varied so greatly.

Clay is highly plastic when wet; it tends to warp or crack when burned. To reduce these tendencies, the colonial brickmaker added sand to clay in a proportion of 30 percent to 70 percent. Sand with rounded grains was preferred to that with angular grains. The mixture of sand and clay would shrink about eight percent in drying and another six percent in firing. Each unit was therefore molded correspondingly larger. During firing, however, there was always some variation in the percentage of shrinkage and some units warped. The finished products had to be sorted to eliminate those bricks that were unsuitable for use and to grade the rest according to quality.

The manufacturing process involved several essential steps, each of which was carried out in a number of different ways. They included:

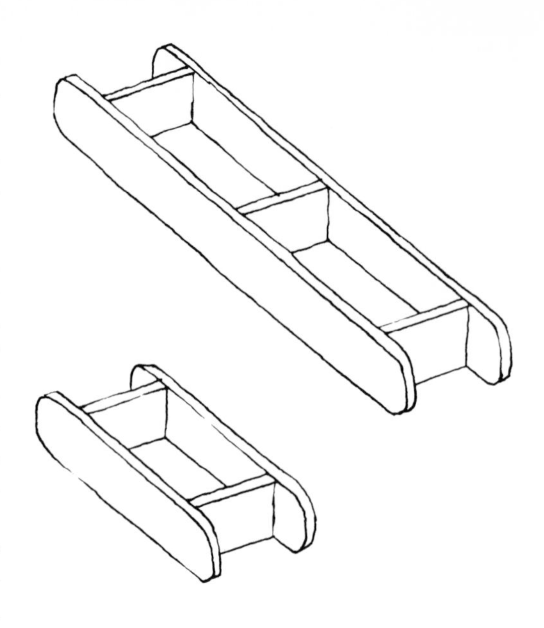

BRICK MOLDS.

Wooden molds were made single or double, with or without bottoms. Some were lined with metal. Cast-iron molds were common in large plants during the 19th century.

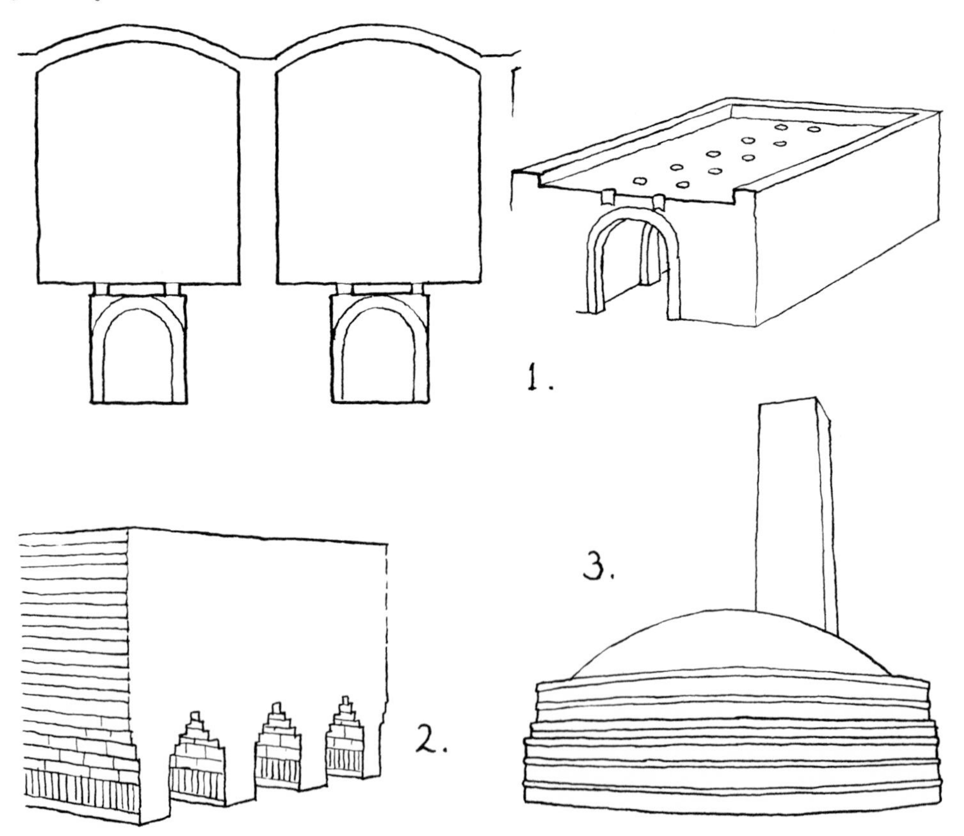

BRICK KILNS.

1. Rectangular kiln. Fuel is burned in the lower compartment and bricks, tiles or pottery are fired in the upper chamber. Variations of this type of kiln have been used since ancient Egyptian times.

2. A clamp or scove-kiln.

3. Round kiln. This type of kiln was common during the 19th century.

Weathering. Spreading the clay mass out in a thin layer to weather for several months or longer. Sometimes, after weathering, the dry clay was crushed in a machine. In careful work, it was screened.

Tempering. Clay, sand and water were mixed in this process. A common method of tempering, although a poor one, was to use a *soak-pit*. Clay, sand and water were dumped into a pit and allowed to soak overnight. This, in itself, accomplished little; when supplemented by spading and treading a satisfactory mixture might be obtained.

Another method of tempering was the use of a *ring-pit*. A heavy wheel traveled in a somewhat irregular circle in a pit, cutting into and mixing the clay. Horses or oxen were driven around the pit to provide power for the wheel.

Pug-mills were introduced into brickmaking in the 19th century, contributing to the mechanization of the industry. A pug-mill usually took the form of an inverted cone or a vertical cylinder, having a series of steel knives or blades fixed on the interior. Between those blades, similar blades were made to revolve on a vertical axis. Clay, sand and water were introduced at the top and thoroughly churned about before being removed at the bottom. The pug-mill was usually driven by steam, but horsepower was sometimes used in small brickyards until the end of the 19th century.

Molding. The wet clay, when about the consistency of soft mud, was placed in wooden or metal molds by hand. Surplus clay was removed from the top with a scraper. After mechanical processes were introduced for molding, this traditional method was called the *soft-mud* process. Sometimes the brick molds had no bottoms; they were placed on a level floor or platform and sprinkled with sand to prevent the clay from adhering to the sides of the molds when they were lifted. Molds with bottoms were more common after the introduction of brickmaking machines. Clay placed in molds by hand was not compressed, so the final product had a relatively low density.

Drying. Clay containing an appreciable amount of water exploded when heated in a kiln; therefore, it had to be thoroughly dried before firing. The moist bricks taken from the molds were set out to dry for a time and turned. When they were strong enough to be stacked a few rows high, they were set on edge with air spaces between them and left in the sun or under roofs, depending on the climate, until they had dried. In humid climates, the drying process was sometimes speeded up by building fires in the drying sheds.

Burning. Burning was the term used for firing or baking. After drying, bricks were stacked for burning. They were heated to a temperature of about 1,800 degrees F. Many bricks were not exposed to that high a temperature and were softer than ideal. Any fuel could be used for burning: brush, straw, wood or coal. Wood was the usual fuel until coal mines were in widespread operation; by the late 19th century, gas-fired kilns set the standard. Many forests had been depleted by the time woodburning of bricks and lime was abandoned.

Several types of *kiln* were employed. Temporary *clamps*, sometimes called *scove kilns*, were the least expensive. They were most often used when bricks were being made at the site of a building but they were also used at commercial brickyards, especially before the 19th century. Before the 20th century, most permanent kilns were round or rectangular in plan; one or more furnaces were fired at the bottom. Smoke and gases from the furnaces passed upward through the air spaces between the stacked raw bricks. Grilles controlled the supply of air.

Clamps were constructed with dry raw bricks. Several walls or *necks* were built, parallel to each other, each about three bricks in thickness. At a height of about two feet, the necks were joined by corbeling courses into a single mass that was built up to a height of eight or ten feet. Throughout the interior, open spaces were left between adjacent bricks, but the outer layers were laid together as closely as possible *(close bolted)*. The tunnels near the bottom served as fireplaces. After fires were started, the ends of the tunnels were blocked off. Sometimes the exterior of a clamp was daubed with mud to reduce the escape of heat. At first the fires were built up gradually, then the heat maintained for several days. After the burning process, it took several days for the bricks to cool. They were then taken out and sorted. Clamps operating in Philadelphia in 1684 fired 40 to 50 thousand bricks at one time, consuming half a cord of wood.[4]

In cylindrical kilns, the raw bricks were often stacked in a radial fashion, one end facing the center; these ends were exposed to the greatest heat and were often burned to a darker color, even to a sort of glazed finish. The ends of bricks closest to the fire in clamps were likewise burned to a darker color than the other faces. Bricklayers made effective use of these *dark headers* to enhance the pattern in the wall. In old kilns, the heat generally was not uniform throughout and variations in hardness and color resulted. Smoke also caused some variation in color.

In some kilns, smoke was conducted through flues around the outer wall, or barriers were placed inside the kiln to prevent gases from the fire from coming in contact with the bricks. This practice was more prevalent in pottery manufacture, in which the value of the product was more impaired by imperfections. During the 19th century, down-draft kilns were gradually adopted; in them, gases from the fire were brought into the firing chamber at the top and withdrawn at the bottom through a flue. More even temperatures were obtained in down-draft kilns.

Types of Bricks. A distinction has long been made between *common* bricks and *stock* bricks, the latter being harder, more regular and more uniform. Stock bricks were placed in the outer face of the wall, because they were better able to resist weathering and gave a more attractive appearance. Toward the end of the 19th century, stock bricks came to be called *facing bricks* or *face bricks* in the United States. Common bricks were used for those parts of the wall that were covered. By selecting the better bricks from each firing of the kiln, stock bricks were separated from the common grade. Some manufacturers made stock and common bricks in different kilns or different firings. Brickmaking was widely distributed throughout the early settlements of the American colonies. Table 6 does not include all the places in which bricks were manufactured.[5]

BRICKMAKING MACHINES

Early Machines. The development of the brick industry in the United States illustrates the gradual introduction of machines to speed up production, improve the product and utilize steam in place of handpower. The earliest patent for a brick-manufacturing machine was granted in 1792; others soon followed. By about 1833 a few bricks were being made by machine but until around 1870 the percentage was small. Machines appear to have gained favor in England sooner than in the United States; indeed, English machines were prominent among types being used in America in the 1870's. At the Centennial Exposition of 1876, in Philadelphia, machines from Canada,

TABLE 6 — EARLY BRICKMAKERS IN THE UNITED STATES

Date	*Place*	*Manufacturer*
1610	Jamestown, Va.	
1629	Salem, Mass.	
1630	Chelsea, Mass.	
c. 1630	Rensselaerwick (near Albany), N. Y.	The Patroon
c. 1630	New Amsterdam (New York), N. Y.	
by 1635	Hartford, Conn.	
by 1638	Maryland	
by 1641	New Haven, Conn.	Atwater
1643	Plymouth, Mass.	
1645	New Haven, Conn.	John Benham
1646	New Haven, Conn.	Edward Shipfield
by 1654	Fort Orange (Albany), N. Y	
1656	New Castle, Del.	
by 1658	New Haven, Conn.	Theophilus Eaton
by 1675	Kent County, Del.	
by 1682	Charleston, S. C.	
by 1682	Philadelphia, Pa.	
1685	Germantown, Pa.	
by 1695	Williamsburg, Va.	Daniel Parke
1701	Williamsburg, Va.	John Tullitt
by 1741	Baltimore, Md.	
mid-1700's	Back River, S. C.	Zachary Villepontoux
by 1758	Pittsburgh, Pa.	

Germany and France were exhibited in addition to those from the United States.

The operations of mixing clay in a pug-mill and pressing the clay in molds demonstrate the first application of machines to brickmaking. Hand-molded bricks were relatively porous. Compressing them by means of a manual screw-press or a power-driven plunger produced denser, stronger and less absorbent bricks. Pressing also insured sharper edges and greater uniformity—qualities that were fashionable during the last half of the 19th century. Machines could handle stiffer clay than was required for hand molding, thus reducing the time for drying raw bricks.

Horizontal and Vertical Types. The design of brickmaking machines was varied. For example, two machines exhibited at the 1876 Centennial illustrate quite different principles.

The machine of Chambers, Brother & Company of Philadelphia was constructed almost entirely of iron and could make 50 to 80 bricks per minute. Clay taken in a crude state from clay banks or a pit was dumped into a hopper; if necessary, sand, loam or coal dust and water were added. From the hopper, the clay passed into a horizontal pug-mill. After being mixed in the pug-mill, the clay was forced through a shaping die by a helical screw. The resulting prism of clay was carried through a cutter, where it was cut into individual bricks. The bricks were carried on an endless conveyor through a sanding chamber and then were placed out to dry.

The forcing of clay through a die sometimes led to the formation of laminations inside the resulting brick. The laminations were planes of weakness. Some machines of the type just described refined the prism of clay by passing it through pairs of rollers before it reached the cutter. Cutting was accomplished by wires or blades that traveled along an oblique path. The angle of the wire or the blade was exactly calculated to compensate for the forward motion of the clay.

The Gregg Brick Machine Company's entry was a circular moldboard machine, working on a vertical axis. It had eight sets of molds, each of which formed four bricks. The crude clay was put in the hopper

> . . . from which, by the action of agitators, it was filled into the molds. The molds, as the board carrying them rotated, passed under a roller and received a steady pressure as they passed. This was a sort of preparatory pressure. The molds then passed under a horizontal knife, placed diagonally, the knife removing from each mold any excess of clay developed by the preliminary pressure. Passing on in their rotary journey, the molds received pressure number two. This was an upward pressure and was caused by a toggle joint. The third and last pressure was a double one, both upward and downward, and was brought about by a simultaneous action of cams and toggles. . . .[6]

The raw bricks were then discharged onto an endless carrier, from which they were removed for drying. The action was entirely automatic. Other machines of this type had molds that passed through them in a straight line. Workmen removed molds that had been filled and fed in

CHAMBERLAIN'S BRICK-MACHINE. At right is a horizontal pug-mill, from which clay is squeezed through a die. The prism of clay is then passed through rollers and cut into raw bricks at left.

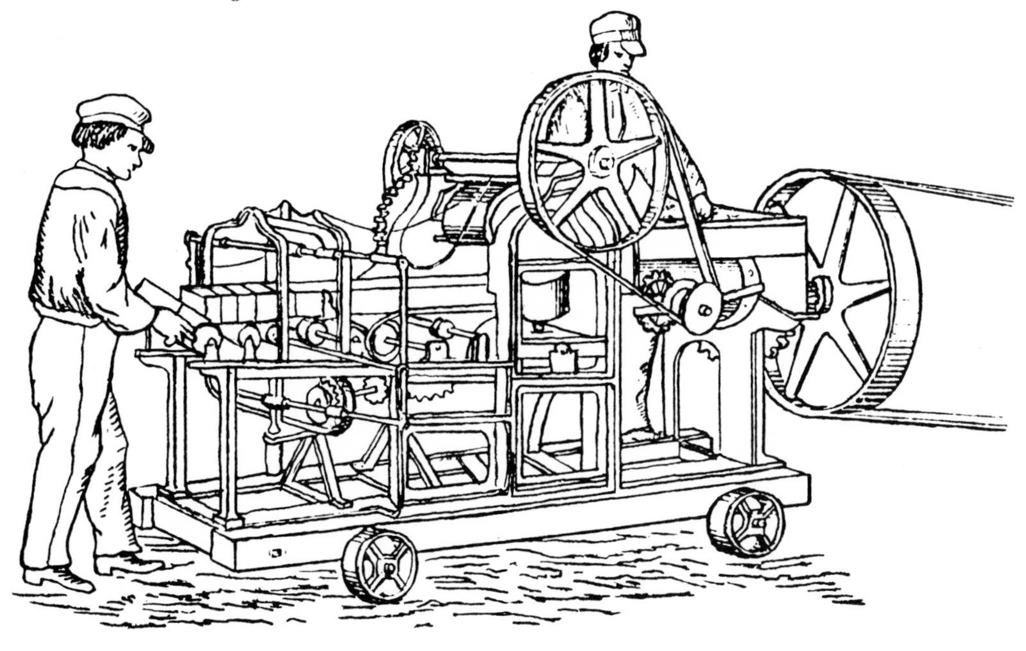

empty ones. Machines such as this appear to have been among the earlier kinds of brickmaking machines. They were well adapted to being run by horsepower and were used in smaller brickyards into the 20th century.

Other types of machine had molds on the periphery of a wheel that revolved on a horizontal axis. Several systems depended on molds carried by a conveyor. One disadvantage of bricks pressed by plungers was their susceptibility to air bubbles. Each different brickmaking process produced bricks that differed in some way from the others. In general, however, pressed bricks were uniform in size, smooth on the surface and durable. A *dry-clay* process was introduced in about 1870 and was widely used in the manufacture of face bricks. Dry clay in the mold was subjected to sufficient pressure to become a cohesive block; these blocks could be baked immediately. The machines were of the plunger type.

Manufacture of Pressed Bricks. Pressed bricks were primarily employed for the exterior faces of walls, and common bricks made by simpler machine processes outnumbered them in production. The common bricks of the late 19th century generally were harder and stronger than their earlier counterparts, although soft bricks were still made by small plants for local consumption.

The manufacture of pressed bricks for facings was carried on in only a few major centers, the best-known ones being Baltimore, Md.; Philadelphia, Pa.; Trenton, N.J.; and Croton, N.Y. The Sayre and Fisher Company was established in 1851 in Sayresville, N. J., by Joseph R. Sayre, Jr., and Peter Fisher.[7] In 1852 this company made three million common bricks. They later began making face bricks by the *stiff-mud* process and in 1863 marketed some that were gray-buff in color. This was one of the early departures from the standard red color.

Another early firm that specialized in pressed face bricks was the Peerless Brick Company of Philadelphia, whose products were shipped all over the country by the 1870's.[8] They began making pressed bricks in hand machines but later switched to power presses and helped to develop the dry-clay process. The term *Philadelphia pressed brick* was so widely used in the 1870's that it was virtually the generic name for that material.

BRICKLAYING

Before the late 19th century, brickwork was distinguished from *masonry* (i.e., stonework) and the *bricklayer* was not called a *mason*. Today, he is commonly called a *brick mason*. As the terminology indicates, the two kinds of workmen were distinct specialists, although most of their

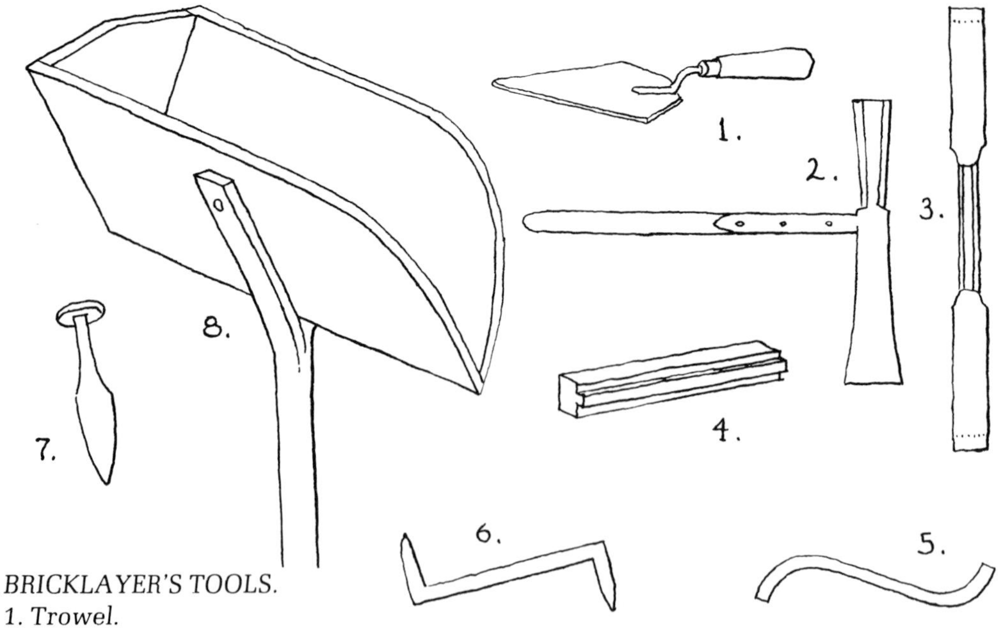

BRICKLAYER'S TOOLS.
1. Trowel.
2. Bricklayer's hammer.
3. Brick-axe.
4. Jointing-rule.
5. Jointer.
6. Raker.
7. Line-pin.
8. Hod for carrying bricks and mortar. Mercer Museum, Doylestown, Pa.

tools and methods were similar. The bricklayer of northern Europe, from whom American practices were derived, was skilled in both routine and ornamental work. He learned his trade by apprenticeship, as did most workmen.

Tools. His tools were a *trowel*, used for spreading mortar and sometimes for breaking bricks; a *bricklayer's hammer*, having one hammerhead and one sharpened peen; a *brick-axe*, having two chisel-like ends that were used in conjunction with a *brick-saw* for cutting and shaping bricks; a *level*; a *plumb-rule* for testing verticals; a *square*; a measuring *rod*; a *compass*; and *lines*. The bricklayer's helper used a *hod* to carry bricks and mortar to where they were needed on the wall. Hoists were also commonly employed to lift these materials. A simple windlass and pulley sufficed for most jobs.

Procedures. When laying walls, the corners of the building were built up several courses, and lines were stretched between them to guide the bricklayer and assure a straight wall with horizontal courses. As the work progressed, the laying was continued all around the building at approximately equal heights, the corners always preceding the portions between them. Differences in level were connected by stepping or *racking* the brick courses. A specification dated October 10, 1839, for Beth Elohim Synagogue, in Charleston, S.C., mentions this requirement.

> Materials for all the foundations walls arches etc. to be of the best quality Carolina Grey Bricks: The pediments architraves to windows, columns, cornices, triglyphs etc. may be of best northern well burn'd bricks. . . . The bricks to be laid three stretchers and one header to be well bedded in soft mortar so that no cavities be left in the walls. . . .no part of the walls shall at the same time be rais'd above the adjacent walls more than four feet, and in all such cases shall be racked back and not united in the progress of the work by toothing, blocks or otherwise.[9]

Traditionally, in good work, bricks of the outer and inner faces of a wall were laid first, followed by the ones in between, one course at a time. Before placing each brick at a face, mortar was spread with the trowel along the bed at the outer edge of the previous course and a *dab* of mortar against the outer vertical edge on the end of the previous brick. The brick being laid was then pressed into place with a sliding motion, forcing the mortar to completely fill each joint. Bricks at the interior of the wall were either *shoved* into their places, onto a bed of mortar, or laid on the bed of mortar and the joints between them *slushed* full of mortar. *Shoved work* was considered the best practice. Sometimes thin mortar was poured in to fill joints that had been left empty by careless workmen; this operation was called *grouting*.

BRICKWORK

Some American traditions had their origins in ancient Roman brickwork. Long thin bricks are still called "Roman." Cutting and carving bricks into ornamental shapes was a Roman practice followed by 18th-century builders in England and America. Most American brickwork before the 19th century was patterned after that used in northern Europe from the late Middle Ages on. The ways in which bricks were laid to bond together in a wall, the sizes of bricks and many characteristic details were adopted by builders in the American colonies and have been used in the United States until the present.

Brick Shapes and Sizes. Although bricks have varied in their exact dimensions and proportions since the late Middle Ages, they have always consisted of small units which a single workman could handle with ease. Larger burned-clay units warp more during firing, so those of small size can be made more uniform and can be laid in a more accurate manner. The shape of a brick that lends itself best to bonding patterns and permits attractive combinations is one in which the length equals twice the width plus one joint width. The length should also equal three times the thickness plus the width of two joints. Variations in brick size are compared in Table 7.[10]

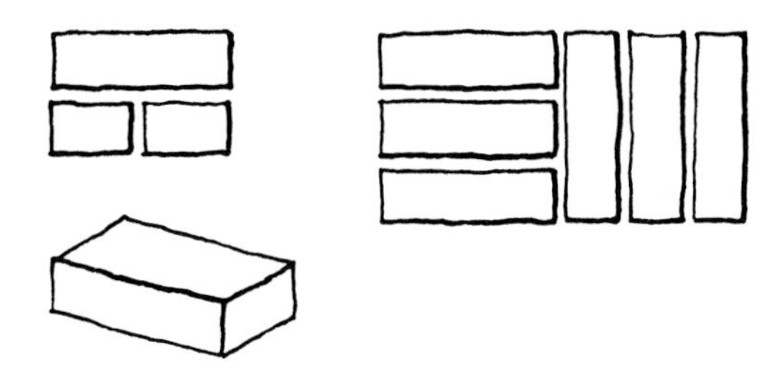

THE IDEAL BRICK SHAPE

Bricks were priced by the thousand (commonly abbreviated as "M"). At various times in England, standard minimum sizes were set by law, to assure purchasers that they would receive at least a minimum amount of material per thousand. Bricks conforming to these regulations were often called *statute bricks*. In the American colonies there do not appear to have been laws governing brick size. Whiffen found no regulations as to size in Virginia, but in some agreements statute bricks were mentioned. "Bricks to be according to the Statute something less than Nine Inches in Length, two Inches and a quarter thick, and four Inches and one quarter Wide" were specified for the construction of St. Peter's Church in New Kent County, Va.[11]

Many bricks made in the colonies conformed to the legal English sizes, whether by intent or by coincidence; they were called "English bricks." This is probably the reason for many claims that bricks for a given building were imported from England. Some bricks were brought in ships as ballast and sold in the colonies. The *Virginia Gazette* recorded shipments in 1737, 1739, 1745, 1753 and 1768 but did not name the sources; the greatest number that arrived in one vessel was 80,000.[12] That quantity of bricks would suffice to construct only one two-story building 20 by 40 feet in size. In 1642, 30,000 "clinker" bricks arrived in a ship at New Amsterdam; however, the purchaser accepted only 10,000 as being fit for use.[13] Bricks made in New Amsterdam were as inexpensive as those imported from across the ocean, so one must be skeptical about stories of imported brick unless the claims can be documented.

Bonding and Bonding Systems. Because bricks are small and relatively light in weight, they must be made to overlap or bond with each other to make a strong wall. They must also be held together by mortar having sufficient cohesive and adhesive strength. Walls are nine, thirteen or more inches in thickness: even multiples of the width of one brick plus one or more joints. Bonding is important not only as an exterior decorative element but as a factor in the strength of a wall. Bonding systems are three-dimensional.

English brick walls before the 15th century were not laid in any definite bond. By the time of American colonization, however, two systematic bonds were in general use in England and were adopted in the colonies. One, called *common bond* in England, was called *English bond* in the colonies. It consisted of alternating courses of *stretchers* (long faces) and *headers* (ends). No matter how thick a wall was, a strong bond was effected without cutting or breaking any bricks, except those near a corner.

Stretcher bond, Hopkins House (c. 1860). Annapolis, Md.

Header bond. Bricks and mortar remain in excellent condition. John Ridout House (1760), Annapolis, Md.

Flemish bond with ruled joints. Palmer House (c. 1754), Williamsburg, Va.

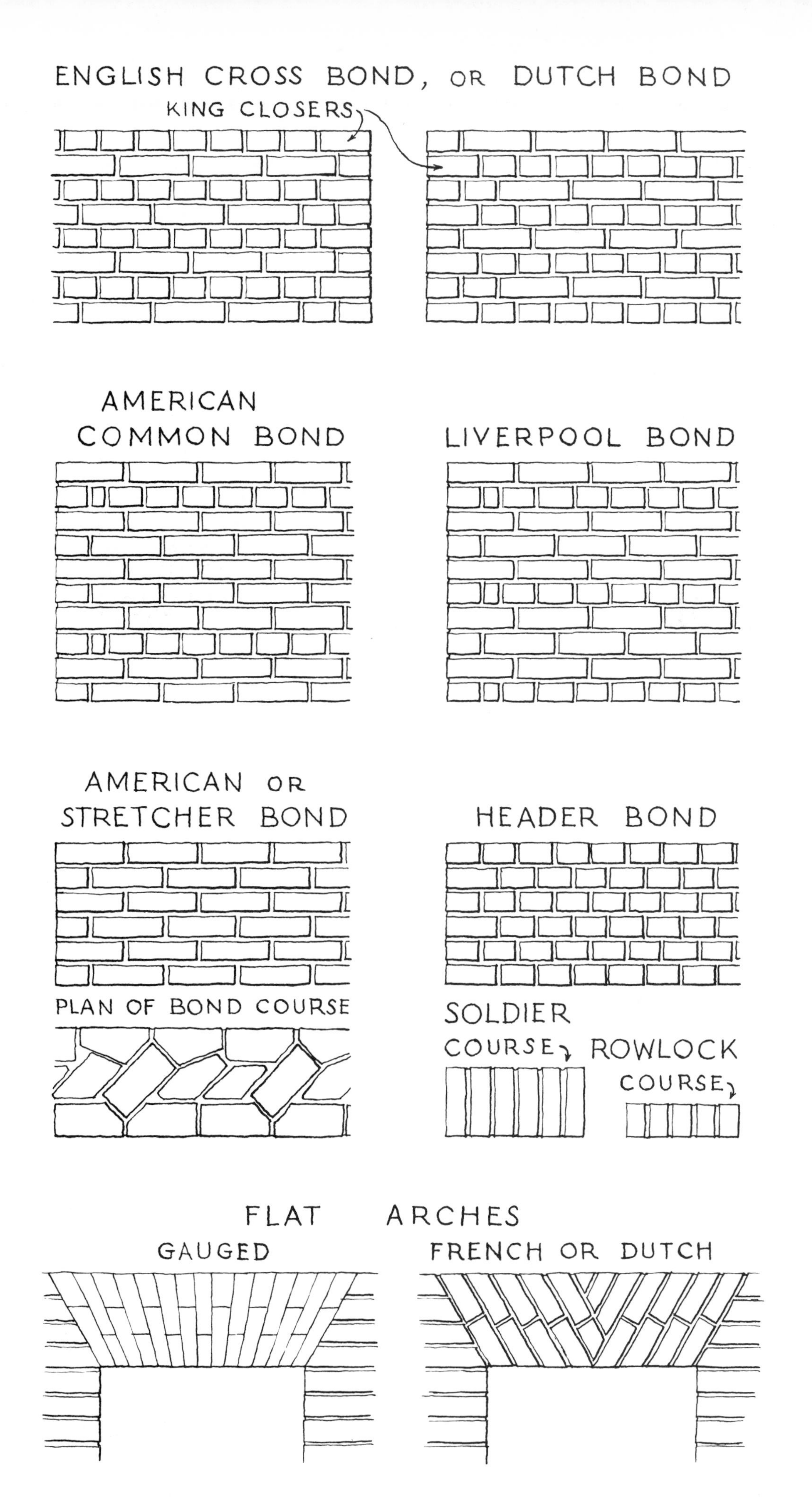
ENGLISH CROSS BOND, OR DUTCH BOND
KING CLOSERS
AMERICAN
COMMON BOND
LIVERPOOL BOND
AMERICAN OR
STRETCHER BOND
HEADER BOND
PLAN OF BOND COURSE
SOLDIER
COURSE
ROWLOCK
COURSE
FLAT ARCHES
GAUGED
FRENCH OR DUTCH

In *Flemish bond* stretchers and headers were alternated in each course. Bricks were arranged so that a header was centered over each stretcher in the course below and vice versa. This system bonded well in depth in a nine-inch wall, but it was time-consuming for thicker walls, because many bricks had to be cut or broken into halves. For each four whole bricks laid in a 13-inch wall, one half-brick was needed as a filler.

Flemish bond became popular in England during the 17th century and is occasionally found in colonial buildings constructed before 1700. A large number of early 18th-century walls in America were built in English bond, but after the middle of the century Flemish bond became more popular and was used well into the 19th century. It is not uncommon to find both bonds used on the same building, especially when English bond was employed for the foundation and Flemish bond for the main stories. Flemish bond was sometimes used on the primary facade of a building while common bond was used for the sides. English bond was used for the thicker foundation walls because of its strength and the ease with which bonding was accomplished.

A variation of the common English bond, which was sometimes called *Liverpool bond*, consisted of the alteration of one header course with three courses of stretchers. It is occasionally found in the United States before the middle of the 18th century and more commonly from then on well into the 19th century. Common *American bond*, in which the bonding course of headers appears once in every six (sometimes five, seven or even eight) courses, was used occasionally in the late

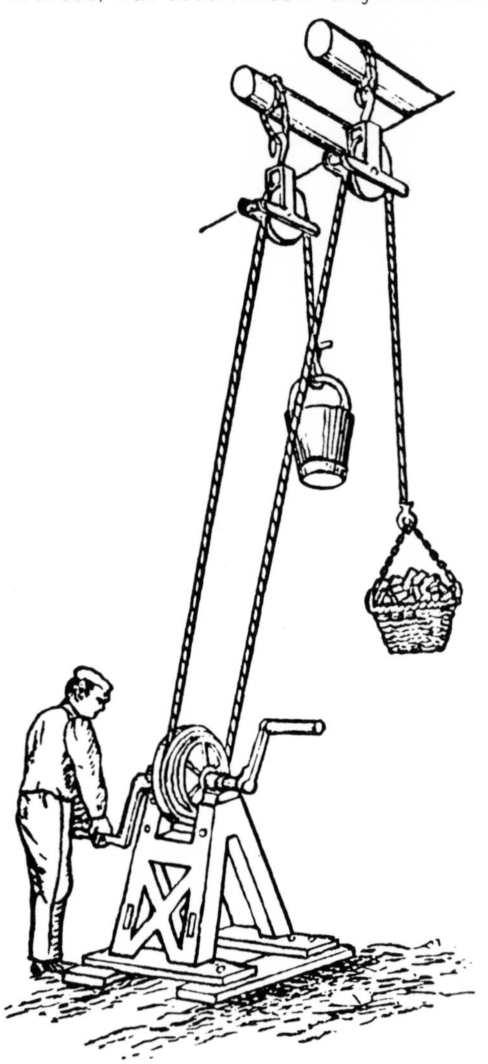

BRICKLAYER'S HOIST. a 19th-century example of an old type: the windlass.

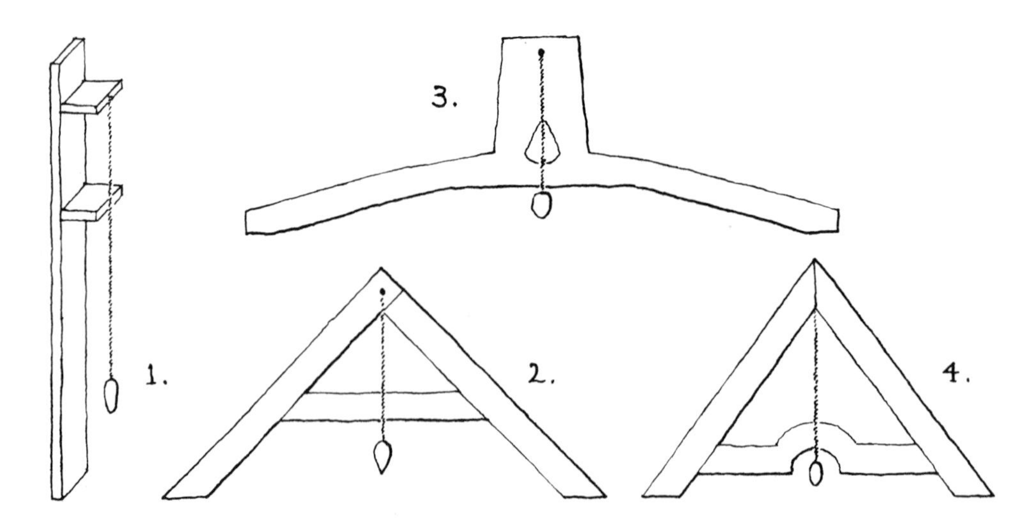

MASONS' LEVELS.
1 -2. Levels employing a plumb line were common from ancient Egyptian times until the 19th century.
3. Medieval European.
4. One French type.

ENGLISH COMMON BOND
HEADER
KING CLOSER
QUEEN CLOSER
STRETCHER
9" WALL
13" WALL

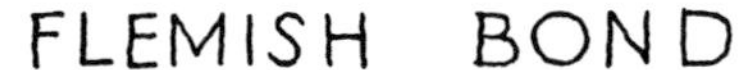
FLEMISH BOND

STRETCHER
KING CLOSER
QUEEN CLOSER
HEADER
9" WALL
13" WALL

18th century and very commonly during the 19th century.

About the middle of the 19th century, *stretcher* or *all-stretcher bond* became fashionable in the United States, particularly for use on the principal facade of a building. The outer bricks were bonded to the interior ones by cutting the inside corners of the face bricks and laying diagonal bonding courses on the interior of the wall at intervals. In spite of the considerable labor involved, the bonding was not very strong.

In English bond, the stretchers were laid so that their end joints lined up vertically. In a variation, called *English cross bond* or *Dutch bond*, the stretchers were placed so that the end joints formed a stepped diagonal line in the masonry of the wall. This bond lent itself to emphasis of diagonal patterning, in which dark headers outlined diamonds or other geometric figures. *Pattern work* of various kinds is found on early American brick buildings but changes in color due to weathering often make the patterns indistinct and difficult to trace today.

All-header, header or *heading bond* was employed in France and England on curved walls; it became fashionable in England in the 18th century.

Heading bond was also in favor in the eighteenth century, not only for circular work . . . but also for whole elevations, as at The Old Hall, Ormsby St. Margaret. One writer regards this as "especially beautiful" [Batty Langley]. Another use of heading bond was when the whole wall surface was to consist of grey bricks. As these were the flared headers and

Terra-cotta trim brought to South Carolina from New York during the 1850's. 101-103 Bull Street, Charleston, S.C.

Molded and rubbed brick trim. West entrance, Falls Church (1769), Falls Church, Va.

as, owing to the method of setting in the kiln, stretchers were not produced this colour, the walling had to be in heading bond as at Wickham and Arundel.[14]

Walls of header bond are rare except in Maryland; many are found in Annapolis and Chestertown.

In addition to considering brick bonds as they are related to the thickness of a wall, it is necessary to examine the exterior corners on both faces, to understand how they were formed. Near a corner, *king closers* (three-quarters of normal brick length) and *queen closers* (split headers, one-half of normal brick width) had to be inserted in order to properly "break joints."

SPECIAL BRICKWORK

The decorative possibilities of standard units of brick have been exploited by builders in several ways: special pattern work, paneling, belt courses and corbeled cornices. Specially made bricks of other shapes were also used to obtain decorative effects. These bricks were fashioned in molds and are called *molded bricks*. They were most commonly employed as *water tables* that gave a pleasing transition from the surface of a protruding foundation to that of the wall above but were also used in decorating other parts of buildings. Columns were sometimes built of pie-shaped bricks; other special shapes were made for angular corners and the curved sections of walls.

Arches, either flat or curved, were frequently built of bricks designed and fired so as to be stronger and more accurate in shape than ordinary ones. They were sometimes molded into a tapered or wedge-shaped form, the angle being appropriate to the radius of the arch. Wedge-shaped bricks were also formed by rubbing rectangular bricks with an

TABLE 7 — BRICK SIZES

	Length	*Width (in inches)*	*Thickness*
Ancient Roman	17-1/2	11-1/2	3-1/4
Ancient Roman	11-1/2	9	3-1/4
Ancient Roman	8	8	1-1/2
Bologna, Italy, 12th century	12-1/2	6-3/8	3-1/8
Flemish, 13th century	8-9-3/4	3-3/4-4-3/4	1-3/4-2-1/2
Dutch, 15th century	6-8-1/2	3-3-3/4	1-3/8-1-3/4
English standard, 1685	8-3/4	4-1/4	2-1/4
English, early 1800's	9	4-1/2	3
Church tower, Jamestown, Va., 1635	8-3/4	4	2-1/4
D. Barnard House, Hartford, Conn. (chimney), 1660	8-1/4	4	2-5/8
Moulthrop House, East Haven, Conn., 1690	6-1/2	2-7/8	1-5/8
Sheldon House, Hartford, Conn., 1715	8	3-7/8-4	2-1/4-2-1/2
Bruton Parish Church, Williamsburg, Va., 1715	9	4-1/2	2-3/4
Hope Lodge, White Marsh, Pa. (chimney), 1723	9-1/8	4-1/4	2-5/8
Stratford, Va. (basement and	8-1/2-9	4	2-1/2
main story), c. 1730	8-3/4	4-1/4	2-1/4
Reynolds Tavern, Annapolis, Md., 1737	8-1/8	3-1/4-4	2-1/2
Brice House, Annapolis, Md., 1760's	9	4	3
Specification for a hospital in Va., by Robert Smith, 1770	8-3/4	4-1/4	2-3/8
Webster House, East Windsor Hill,	7-3/4	3-7/8	2
Conn., 1787	10-1/2	4-3/8	2-1/2
Tallman Warehouse, New Bedford, Mass., 1790's	6-3/8-7	3-3/8-3-1/2	1-1/2-1-3/4
Blacklock House, Charleston, S.C., 1800	9	4-3/8	2-3/4
Blacklock Gazebo	9-1/4-9-1/2	4-3/8	3
Moore House, Rochester, N.Y., 1831	7-1/2	3-5/8	2-1/8
Hopkins House, Annapolis, Md., 1860's	8	4	2-1/8
National Brickmakers' Assoc., standards adopted 1899:			
common bricks	8-1/4	4	2-1/4
face bricks	8-3/8	4	2-3/8
Roman bricks	12	4	1-1/2

abrasive block (*rubbing stone*). Bricks for window and door jambs, quoins and other details were often formed by the same rubbing process; the smooth, rubbed faces contrasted in color with the rest of the wall. Rubbed bricks were sometimes laid with thinner joints than other bricks in a wall.

Carved or *cut* brickwork was used sparingly in the United States. *Cutting bricks* were soft and fine textured; they were made with a high percentage of sand in the clay. Carving was done with chisels and rasps. *Rubbing bricks* were nearly similar in composition but harder than cutting bricks.

TERRA-COTTA

Architectural terra-cotta is closely related to brick but is finer in texture. For ornamental work, it was shaped in a mold or modeled by hand before being fired in a kiln. Large pieces of clay are subject to warping in the kiln, so great care was exercised in the manufacture of terra-cotta.

Origins. Terra-cotta is an ancient building material. It was used to face crude brick and timber structures by the ancient Greeks and Etruscans. During the late Middle Ages and the Renaissance it was used in Italy to create some of the best architectural ornament and relief sculpture ever made. From Italy, some workmen carried the methods of working in terra-cotta to northern European countries.

During the late 18th century, several manufacturers in England and France developed materials of secret composition that were somewhat comparable to terra-cotta. These building materials were used in architectural ornaments and trim. Probably the best known of these products was "Coade stone"; it was used in the construction of several mantelpieces in the Octagon House (c. 1800) in Washington, D.C.

Terra-cotta proper appears to have been little used in Europe after the Renaissance, until around the middle of the 19th century. The history of its revival is rather obscure. In England, the reintroduction of terra-cotta for architectural purposes on a major scale took place in 1842–44, on a church built at Lever Bridge, near Bolton. The architect of the church was Edmund Sharpe. Another important building on which terra-cotta was used was Albert Hall, a concert hall built in London in 1867–71. The terra-cotta work on these buildings was not as true in line or in color as that produced by the end of the century, but it was not seriously affected by weathering. The entire exterior of the Natural History Museum in London, built in 1878 from the design of Alfred Waterhouse, was of terra-cotta.[15]

Early Use in the United States. Some ornamental terra-cotta was used in the United States by the middle of the 19th century. Terra-cotta keystones in the form of tigers' heads decorated Richard Upjohn's Trinity Building, built in New York City in 1851–52. The Hotel St. Denis, built in 1853 at Broadway and 11th Street in New York City, had elaborate window trim, string courses and a cornice of locally manufactured terra-cotta; the architect was James Renwick, Jr. By 1896, although the terra-cotta on both of these buildings had been painted over, there was no evidence of decay of the material.[16]

The Boston Museum of Fine Arts (c. 1872), designed by the firm of John Hubbard Sturgis and Charles Brigham, had terra-cotta trim and ornamental panels. This material was imported from Stamford, Lincolnshire, England. The museum attracted much favorable attention to architectural terra-cotta in the United States.[17]

One of the early American dealers in terra-cotta was Ambrose Tellier of 1194 Broadway in New York City. Tellier is listed in an 1859 business directory. Louis Scharf and William Gilinger began making terra-cotta in Whitemarsh Township, Montgomery County, Pa., in 1856.

Terra-cotta frieze and keystone; special brick moldings. Pension Building (1882-86), Washington, D.C. Montgomery C. Meigs, architect; Casper Buberl, sculptor. (Jack E. Boucher for Historic American Building Survey.)

They first ground and worked the clay by hand but later substituted horsepower. Moorehead's Terra-Cotta Works was established in the same county in 1866.[18]

Alfred Hall adapted part of his brick plant at Perth Amboy, N. J., for making plain and ornamental terra-cotta blocks in about 1876. This department later developed into the Perth-Amboy Terra-Cotta Company.[19]

There is little distinction between molded bricks and small pieces of terra-cotta; enriched molded bricks were sometimes made by placing clay into molds and compressing them in hand-operated screw presses. The natural reddish color of terra-cota was similar to that of face bricks. When another color was desired, a "slip" or wash of clay was applied to the surface. During the mid-19th century raw blocks of terra-cotta were often touched up with sandpaper before being fired in the kiln. Touched-up places did not withstand weathering well. "The method in vogue at that time, but long since abandoned, of rubbing down irregularities with coarse sandpaper, when the work was white-hard, and just previous to burning, has much to answer for. It disturbed the plastic surface that had formed on the face of the block, in obedience to the laws of molecular cohesion, leaving such parts tender and porous. . . . every indication of fracture or decay is found on blocks that have been sandpapered before burning.[20]

During the 1890's, two circumstances favored wide acceptance of both plain and ornamental terra-cotta in American architecture. The rapid development of steel-framed tall buildings created a demand for lightweight fire-resistant materials for covering. At the same time, Italian Renaissance architecture was widely admired and many American architects were proficient in adapting its elements to the design of modern buildings. Polychrome terra-cotta ornament modeled after the Italian precedent became an important part of American architecture and its manufacture flourished well into the 1920's. Other ornaments were successfully applied to terra-cotta during this period. The personal style of the architect Louis Henri Sullivan reflected some of the best terra-cotta of the period.

DETERIORATION

Causes. The majority of bricks used in America during the 18th and 19th centuries were soft and porous by present-day standards. They absorbed 20 to 25 percent of their weight in water, whereas 10 percent or less was considered the accepted maximum by the end of the 19th century. Soft, underburned bricks might even absorb as much as 35 percent of their weight in water. The absorbency factor is important to bear in mind when comparing modern bricks and old ones.

The action of moisture on a brick wall is similar to that described in the chapter on stone: Rising damp, rainwater and condensation carry dissolved acids that attack the material. Soluble salts crystallize in the pores of a brick or on the outer surface. Alternate freezing and thawing bring about deterioration. Dirt and airborne particles accelerate damage to the exterior surface. Naturally, the climate in which a building is located is a major factor in determining the kinds and extent of the processes of deterioration.

It was not uncommon for new brick walls to develop *efflorescence.* Soluble salts found in the brick or mortar, or formed by interaction between the two, reached the surface of the brick and dried out there. Eventually these salts were removed by natural action or by brushing and washing. Groundwater, rising by

TERRA-COTTA WINDOW. Hotel St. Denis (1853), New York, N.Y.

capillary action, also introduced harmful salts. As the salts became concentrated in the lower parts of the wall the dampness rose even higher. This action sometimes caused the face of bricks to disintegrate.

Leaking roofs, gutters and parapets also constituted a major source of water in walls. Even small cracks where mortar failed to adhere to a brick introduced water.

The outer crust of each brick is harder and more dense than the material inside. Once this crust is removed by any means—rising damp, freezing and thawing or sand blasting—disintegration of the brick is greatly accelerated.

Old bricks frequently develop cracks where shrinkage or laminating occurred in the clay or where unequal stresses were set up during firing. The corners of a brick commonly break or wear away more than the rest of the face, giving it a somewhat rounded surface. Repointing of the joints is the most common operation in maintaining and repairing a brick wall; if this is improperly done it contributes to deterioration of the bricks. Repointing is discussed in the chapter on mortar.

Remedial Measures. Once a brick begins to crumble, crumbling invariably continues and the condition cannot be remedied except by replacing the unit. A few bricks can be removed and replaced with sound bricks without endangering a wall. A dampproofing course can be introduced into a wall in short lengths by removing a few bricks at a time and inserting a waterproof membrane in the joint before the bricks are replaced.

It is difficult to make repairs that closely match old work. Weathering often helps to harmonize new work with untouched portions of a wall but it can, at times, produce the opposite effect. In addition, repairs made by persons who lack the necessary knowledge of old materials or who do not possess the skill and patience to execute the repairs in a proper manner often harm the physical structure of a building as well as its historic character. If major structural repairs are made, improper methods or poor workmanship may even endanger lives.

Weathered bricks. Some fissures and worn corners; at lower left, chipping below a nail hole. Palmer House (c. 1754), Williamsburg, Va.

Flemish bond with eight-inch arch over door constructed of ordinary bricks. Marks of pick and point are visible on the stone steps and foundation. Fox Stand (Inn), Royalton vicinity, Windsor County, Vt. (Jack E. Boucher for Historic American Buildings Survey.)

Ornamental chimney pots. Bishop House (c. 1850), New Brunswick, N.J. (Jack E. Boucher for Historic American Buildings Survey.)

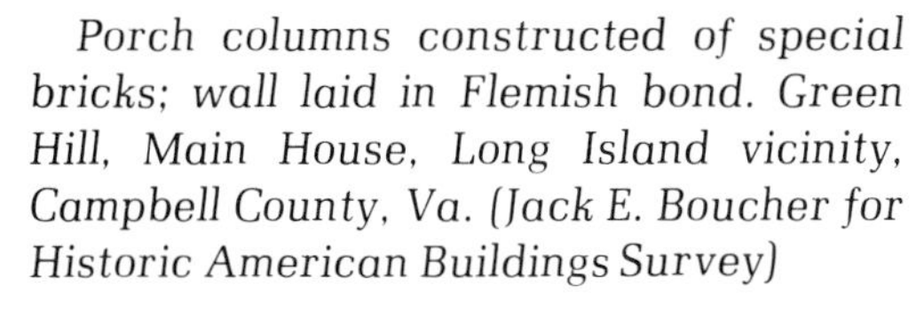

Porch columns constructed of special bricks; wall laid in Flemish bond. Green Hill, Main House, Long Island vicinity, Campbell County, Va. (Jack E. Boucher for Historic American Buildings Survey)

Ornamental brick trim. This was originally to be covered with plaster. St. Luke's Episcopal (now Fourth Baptist) Church (1859), Charleston, S.C. Francis D. Lee, architect.

Molded brick water table, header bond. James Brice House (c. 1760), Annapolis, Md.

Gauged brick arch. Wythe House (c. 1754), Williamsburg, Va.

FOOTNOTES FOR CHAPTER II

1. Marcus Whiffen, *The Public Buildings of Williamsburg* (Williamsburg, Va.: Colonial Williamsburg, 1958), p. 21; Marcus Whiffen, *The Eighteenth-Century Houses of Williamsburg* (Williamsburg, Va.: Colonial Williamsburg, 1960), p. 6.
2. Brian Mason, *Principles of Geochemistry*, 2nd ed. (New York: Wiley, 1960), pp. 151–54.
3. Heinrich Ries, "Report on the Clays of Maryland," *Maryland Geological Survey*, vol. 4 (Baltimore: Johns Hopkins Press, 1902), p. 301. *See also Clays of New York*, Bulletin of the New York State Museum, vol. 7, no. 35, (Albany, N.Y.: The Museum, 1900), for a description of brickmaking.
4. J. Thomas Scharf and Thompson Westcott, *History of Philadelphia*, vol. 3 (Philadelphia: L. H. Everts & Co., 1884), p. 2293, citing Jonathan Dickinson in the Logan Papers.
5. Compiled by the author from many sources.
6. J. S. Ingram, *The Centennial Exposition Described and Illustrated* (Philadelphia: Hubbard Bros., 1876), pp. 198–200.
7. *The Brickbuilder*, vol. 6, no. 2 (February 1897), p. 42.
8. *The Brickbuilder*, vol. 4, no. 10 (October 1895), p. 203. Letter to the editor by Geo. M. Fiske.
9. Transcribed by the author by courtesy of Thomas J. Tobias, a descendant of the builder, David Lopez. Photostat copies of this contract may be obtained from the American Jewish Archives, 3101 Clifton Avenue, Cincinnati, Ohio 45200.
10. Compiled by the author from many sources; some measurements by the author.
11. Whiffen, *The Eighteenth-Century Houses*, p. 6.
12. Ibid., p. 7.
13. W. J. A. Arntz, "Export van Nederlandsche Baksteen in Vroegere Eeuln, Medegedeeld Door," *Economisch-Historisch Jaarboek* (The Hague: Nijhoff, 1947), chap. 2. Courtesy of Theodore H. M. Prudon.
14. Nathaniel Lloyd, *A History of English Brickwork* (London: Montgomery, 1925), p. 66.
15. Thomas Cusack, series of articles on architectural terra-cotta in *The Brickbuilder*, beginning in 1896.
16. *The Brickbuilder*, vol. 5, no. 12 (December 1896), p. 227.
17. Ibid., p. 229.
18. Theodore W. Bean, ed., *History of Montgomery County, Pennsylvania* (Philadelphia: Everts & Peck, 1884), p. 626.
19. *The Brickbuilder*, vol. 5, no. 12 (December, 1896), p. 227.
20. Ibid., p. 229.

III. MORTAR

INTRODUCTION

Mortar performs several essential functions. The plastic nature of fresh mortar permits it to fill voids between masonry units that do not exactly fit together and to gradually become adjusted to movements within a wall that occur during construction. Mortar helps make walls watertight; it enables bricks and small stones to form a coherent mass and acts as a lubricant to permit the sliding of heavy stones into position. The form and color of mortar joints contribute significantly to the appearance of a wall.

It is essential to distinguish between *hard* and *soft* mortars. The use of lime-sand mortar predominated until about 1880. It was soft enough to furnish a plastic cushion that allowed bricks or stones some movement relative to each other. The entire structural system depended upon some flexibility in the masonry components of a building. A cushion of soft mortar furnished sufficient flexibility to compensate for uneven settlement of foundations, walls, piers and arches; gradual adjustment over a period of months or years was possible. In a structure that lacks flexibility, stones and bricks break, mortar joints open and serious damage results.

Cement mortars used after about 1880 are hard. Joints of cement mortar are strong and unyielding; they are appropriate to modern bricks and concrete blocks. Rigidity is characteristic of modern masonry. Those accustomed to cement mortar and concrete must consider the elasticity of historical masonry when restoring historic buildings. Hard and soft building materials cannot be used together effectively.

Several properties must be considered when judging the appropriateness of a particular mortar: cohesiveness, adhesiveness, strength, setting time, hardening time, handling ease, ability to set and harden under water (*hydraulic* quality) and the degrees of expansion and solubility. The visual aspects of color and texture are also of significance.

CLAY MORTAR

Clay was the first material to be used for mortar. It has been used throughout history in walls of unburned brick. In America, clay mortar was used for ordinary brick and stone walls in regions where lime was difficult to obtain. Often, clay was used because of its low cost. Clay mortar joints can bear heavy loads, but in humid climates they need protection from rain; it is thought that pent roofs at the second-floor level of stone houses in eastern Pennsylvania originated to serve this purpose. In his diary, Judge Samuel Sewall recorded a disaster which occurred in Massachusetts: "October 30, 1630, a stone house which the governor was erecting at Mystic was washed down to the ground in a violent storm, the walls being laid in clay instead of lime."[1]

The earliest walls of field stone and split gneiss in Connecticut were laid dry or with clay mortar. In northern New Jersey and in the Hudson Valley, mud and clay strengthened by straw or hogs' hair was used for mortar as late as about 1790. New Jersey clay was relatively impervious to moisture but it could be eroded by strong rains. Many masonry houses there were repointed with lime mortar.[2] Clay mortar was used in other stone-building areas during colonial times.

A standard specification for building lock keepers' houses on the Chesapeake and Ohio Canal, written about 1828, required stone walls to be laid with clay mortar "excepting 3 inches on the outside of the walls above ground and the inside of the cellar which 3 inches to be good lime mortar and well pointed."[3]

Old interior chimneys are commonly found to have been constructed with clay mortar below the roof line; above that level they were pointed with lime mortar. Clay mortar in chimney flues was baked quite hard by the heat. Clay was used as mortar in the chimney of the Godfrey Nims House (1718) in Deerfield, Mass. The roof was raised in 1785 but the earlier roof line is revealed by the point at which the use of clay mortar was terminated.[4]

GYPSUM MORTAR

The earliest known use of gypsum mortar was in ancient Egypt, where mortar of fluid consistency was poured into narrow joints to fill them and used to lubricate the bed of large stones while they were being moved into position. Gypsum mortar has been used in Persia from ancient times to the present. In medieval France gypsum mortar was used to bond small stone members when they were required to support unusually heavy loads. In the United States, gypsum was sometimes used to set marble trim and tiles and as a minor ingredient in lime and cement mortar. The use of gypsum for plastering is described in the chapter on plaster.

LIME-SAND MORTAR

Lime-sand mortar was the most common type used in structures located above water level until the late 19th century. At that time its popularity declined, but significant quantities of lime mortar are still used. It is of great importance in the repair and restoration of historic buildings.

Lime was obtained chiefly from limestone and marble. Shells were frequently used as a source of lime in regions that were deficient in these materials or regions where deposits of these rocks were not worked. Along the Atlantic coast of the United States, the colonists often found shells in Indian middens. They burned the shells for lime and sometimes crushed them for gravel. In 1688 John Clayton wrote a letter to the Royal Society in which he described piles of discarded shells in Virginia: "In some Places for several Miles together, the Earth is so intermix'd with Oyster shells, that there may seem as many Shells as Earth; and how deep they lie thus intermingled, I think is not yet known. . . . In several Places these Shells are much closer, and being petrified, seem to make a Vein of Rock. . . . Of these Rocks of Oyster-Shells that are not so much petrified they burn and make all their Lime; whereof they have that store, that no Generation will consume."[5] Such shell deposits were also found in Connecticut. "In the southern part of the state, and especially in towns situated near the Sound, where oyster shells were easily obtained and burned to produce lime, the use of regular mortar in the construction of masonry was earlier and more common. . . . Lime was used very early in New London and New Haven, and the records of the latter town for November 3, 1639, refer to it. . . . The early court records of New Haven repeatedly refer to the 'oyster shell field,' which was situated east of State Street, between Chapel and George. It was probably an Indian refuse heap. . . ."[6]

Shell fragments and bits of broken Indian pottery have been observed in mortar of the late 18th century at Charleston, S.C.[7]

Because limestone is of sedimentary origin, its composition varies with the circumstances of sedimentation. Mud, clay and sand are common impurities. Pure limestone is calcium carbonate; as quarried, limestone may contain alumina, silica, magnesia or other substances. Magnesian limestone makes good building stone but the value of mortar made with lime from magnesian stones is debatable. Before the development of scientific methods of analysis, producers of lime had to judge the quality of their sources by experience.

LIME BURNING

When calcium carbonate is heated to about 1650 degrees F., carbon dioxide is given off. The remaining calcium oxide is called *quicklime*.

Kilns. The heating, "burning" or, more

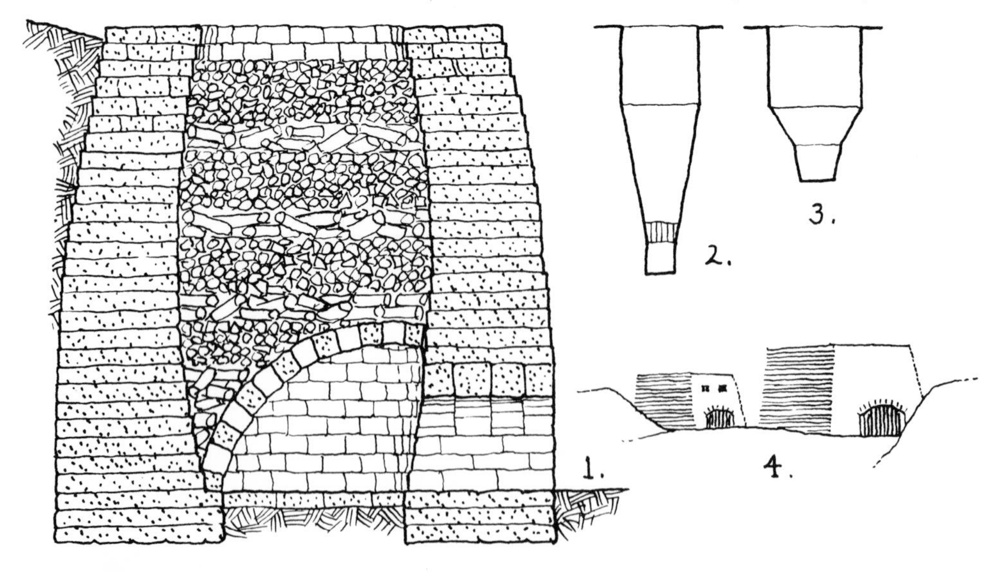

LIME KILNS.
1. Common type, built into a hillside. Often lined with firebricks in the 19th century. Sometimes fuel and lumps of limestone were placed in alternate layers, as shown here; sometimes fuel was burned only at the bottom.
2-3. Variations in shapes and proportions of the common lime kiln.
4. Kilns near Spring Mill, Pa., 1852.

properly, *calcining* of limestone (commonly called *lime rock*) was sometimes done over open fires but kilns were more often employed. Some limestone was burned on large heaps of logs during the fall and winter of 1818–19 for the construction of the Erie Canal.[8] The traditional method of lime burning in Yucatan was comparable:

> In the smaller towns and villages of northern Yucatan, lime-kilns are still made as they were in ancient times, and the local limestone is still burned to make lime just as it probably has been done for the last two thousand years. A place in the forest is selected and completely cleared. Fagots of wood about two feet long are next cut and laid in a circle varying from ten to twenty feet in diameter depending on the size of the kiln to be built. These fagots are laid with their lengths parallel to the radii of the circle, a hole about a foot in diameter being left in the center. This neatly laid pile of wood is built to a height of about four feet. On top of it, beginning about a foot back from its outer edge, are piled broken pieces of limestone about the size of one's fist. These are heaped to a height of another two feet.
>
> When this is finished the kiln is fired by dropping leaves and rotten wood into the bottom of the hole at the center and igniting them. The fire thus works from the bottom up and from the inside of the kiln outward. . . . It takes thirty-six hours for a kiln to burn completely, and when a good burn has been achieved the limestone fragments are completely reduced to a pile of powdered lime.[9]

Permanent kilns were made of stone or brick.[10] When the terrain permitted, they were built into a hillside to facilitate loading from the top. Types built in America were similar to European models of the late Middle Ages to the 19th century. They were about 20 feet high, round or polygonal in plan and 10 feet or more in internal diameter; some had barrel-shaped interiors. At the bottom of the kiln there was an opening through which the fire could be ignited and the burned lime could be removed. This opening could be closed to control the draft. Kilns were open at the top.

Before loading a kiln, a crude arch of stones was constructed at the bottom, inside the opening. Alternate layers of firewood and lumps of limestone were then loaded from the top until the kiln was full; then it was ignited from the bottom. The rate of burning was controlled by adjusting the flow of air into the opening at the bottom. After burning for one and a half to two days and cooling for an equal length of time, *quicklime* was taken out from the bottom of the kiln. Ashes and improperly burned pieces were removed. Quicklime retained the lumpy form of the stones from which it was made.

Fuel Consumption. Lime burning consumed a considerable amount of fuel; about 60 cubic feet of oak, 117 cubic feet of fir or 450 pounds of coal were needed to produce one ton of lime.[11] An appreciable saving in fuel was effected by continuous operation of a kiln; this method was followed by many major producers during the 19th century. As the materials in a kiln settled, more fuel and limestone were added in alternate layers and lime was periodically taken out from the bottom.

Care had to be taken in the handling of lime to maintain the quality of the material. Some builders preferred to use lime as soon as possible after it was removed from the kiln. When quicklime is exposed to air or moisture, portions of it become inert and unfit for use in making mortar. That which was shipped or kept in storage was put into tightly sealed barrels.

For making mortar, quicklime (often called *unslaked* lime) was sometimes used. At other times quicklime was first *slaked* or *hydrated*. Both methods are described in this chapter.

SLAKING LIME

When water is added to quicklime (calcium oxide), slaked or hydrated lime (calcium hydrate) is formed. Heat is given off during slaking. Slaked lime is a powder whose volume is greater than the volume of the quicklime from which it is made. There are four methods of slaking lime:

(1) *Sprinkling* or *drowning*. The correct amount of water was sprinkled onto quicklime. The lumps cracked open and dry powder was formed.

(2) *Immersion*. Quicklime was placed in a basket, lowered into water and drawn up in time to complete the slaking action in the air. The correct time of immersion was a critical factor which was difficult to determine.

(3) *Exposure*. Quicklime was simply exposed to the air in a shed or shelter for a considerable length of time. It absorbed moisture from the air and became partially slaked, but it also absorbed carbon dioxide and thus acquired inert material (calcium carbonate) that adulterated the slaked lime. This method was universally considered the least satisfactory one.

(4) *Making lime paste*. Quicklime was placed in a pit or a vat and more water than the amount required for slaking was poured over it; the mixture was allowed to stand and slake. This lime paste was either used at once or stored in a covered pit for months or years.

When quicklime is slaked, the increase in volume varies according to the kind and amount of impurities that are present. Therefore, the quantity of quicklime needed to make mortar of a desired quality varied; it was necessary to know the properties of the lime from each source in order to establish the correct ratio of lime to sand and water. Mortar formulae given in old builders' handbooks must be studied with caution if the kind or source of lime is not specified.

Fat limes absorb about one-half their volume in water during slaking and increase greatly in volume. Mortar made from fat lime was preferred by masons because of its "rich" or "oily" consistency. *Meagre* or *poor limes* absorb less water, give off less heat during slaking and yield a smaller quantity of mortar. However, some meagre limes impart desirable properties to mortar that make it useful in damp locations. Table 8[12] illustrates the degree to which lime from different sources changes in volume during slaking.

COMMON LIME MORTAR

Ingredients. The common variety of mortar was made of lime, sand and water. Details of its preparation varied according to regional customs and individual preferences but most of these details were well known throughout Europe and America. The builder was aware of more methods than he practiced.

Mortar was prepared for use by the following basic methods:

(1) Dry pulverized quicklime and dry sand were mixed. Water was then added and the whole mass was mixed.

(2) Dry sand was added to lime paste and thoroughly mixed in. If necessary, water was added.

(3) Slaked lime power, sand and water were mixed together, either simultaneously or by adding water to previously mixed lime and sand.

Proportions. Examples of these general methods and modifications of them follow. The descriptions include the relative proportions of lime to sand.

> Sand is added to lime for economy and to prevent shrinkage. Sand should be . . . in such quantity that the lime will fill all the interstices. If an excess of sand is used, the bond is poor. If too little sand is used, the mortar shrinks and cracks. If too little lime is used the paste is made thin. In ordinary sands, the spaces form 39% to 40% of the total volume, and in such 1 vol. paste fills voids of 2½ vol. sand. In practice 1.25 to 2 vol. of sand to 1 of paste is used. This in case of fat lime means 3 to 5 vol. of sand to 1 measured vol. of lime. This gives a plastic mortar which does not crack.[13]

Vitruvius was an ancient Roman architect whose books were widely quoted and whose precepts were highly respected in Europe and America. He recommended the following proportions: ". . . mix your mortar, if using pitsand, in the proportions of three parts of sand to one of lime; if using river or sea-sand, mix two parts of sand with one of lime. . . . Further, in using river or sea-sand, the addition of a third part composed of burnt brick, pounded up and sifted, will make your mortar of a better composition to use."[14]

Loriot was a French engineer of the 18th century whose knowledge of mortar was widely acclaimed by engineers in England and America. His formula was made public in 1774: ". . . take one part of brick dust finely sifted, two parts of fine river-sand skreened, and as much old slaked lime as may be sufficient to form mortar with water in the usual method, but so wet as to serve for the slaking of as much powdered quick-lime, as amounts to one-fourth of the whole quantity of brick-dust and sand. When the materials are well mixed, employ the composition quickly, as the least delay may render the application of it imperfect or impossible."[15]

Clay and pulverized bricks or shells were sometimes added to early American mortars. Norman M. Isham and Albert F.

TABLE 8 — CHANGES IN LIME VOLUME DURING SLAKING

	Fat lime of Strasbourg	*Yellow lime of Obernai*	*Alt kirch lime*	*(meagre) "Boulogne pebbles"*
1. Volume of fresh (quick-) lime	1	1	1	1
2. Water to slake it to a dry powder	0.5	0.5	0.5	0.33
3. Volume of dry powder (slaked lime)	3.5	2.0	1.63	1.16
4. Additional water to form a wet paste	1.5	0.25	0.13	0.17
5. Volume of wet paste (lime paste)	1.75	1	1	0.75

Brown described this practice in *Early Rhode Island Houses.* "The earliest mortar of the colony is what is called shell mortar. Perhaps the best known specimen of masonry built with this material is the stone mill at Newport, built by Governor Benedict Arnold somewhere about 1670. This mortar is identical with that of the Bull house and of some other buildings. It is composed of 'pulverized shells, clay, sharp sand, and fine gravel.' This sort of cementing material was used in the other settlements of the colony and seems to have lingered until quite late. . . ."[16]

George Mason wanted no clay added to the mortar of his home, Gunston Hall, built in Fairfax County, Va., in 1758.

> When I built my House I was at pains to measure all the Lime and Sand as my Mortar was made up and always had two Beds, one for outside-work ⅔ Lime and ⅓ Sand. . .it is easily measured in any Tub or Barrel. . .If you have any good pit sand, out of your Cellars or well, it will make your mortar much tougher and stronger. . . .Next to pit sand the River Shoar Sand on fresh water is best and the Sand in the road worst of all; as being very foul and full of Dust.
> I wou'd by no means put any Clay or Loam in any of the Mortar, in the first place the mortar is not near so strong and besides from its being of a more soft and crumbly nature, it is very apt to nourish and harbour those pernicious little vernim the Cockroaches. . . .[17]

Mixing. Thorough mixing or *beating* was emphasized by most authorities. In 1823, Peter Nicholson recommended a practice in *The New Practical Builder* which was probably well agreed upon in England and the United States.

> Before the mortar is used, it should be beaten three or four times over, so as to incorporate the lime and sand, and to reduce all knobs or knots of lime that may have passed the seive. This very much improves the smoothness of the lime, and, by driving air into its pores, will make the mortar stronger: as little water is to be used in this process as possible. Whenever mortar is suffered to stand any time before used, it should be beaten again, so as to give it tenacity, and prevent labour to the bricklayer. In dry hot summer-weather use your mortar soft; in winter, rather stiff. If laying bricks in dry weather . . . wet your bricks by dipping them in water, or by causing water to be thrown over them before they are used. . . .[18]

SETTING AND HARDENING OF COMMON LIME MORTAR

After being mixed, common lime mortar remains plastic for several hours. It must be placed in the wall while in this condition. Mortar is said to have *set* when it loses its plasticity. When set, it will support the load of masonry placed on it if the work does not proceed too rapidly. After setting, mortar hardens at a slow rate, taking months or years to attain its ultimate strength. Sometimes common lime mortar on the interior of thick walls never really hardens.

Mortar sets and hardens approximately as follows: As mortar dries, some of the hydrated lime crystallizes and binds the mass together; this constitutes setting. Hardening takes place largely when carbon dioxide from the air combines with hydrated lime in the mortar to form calcium carbonate. This does not occur if the mortar is either too wet (containing more than five percent water) or too dry (containing less than seven-tenths of one percent water). Mortar at the surface of a joint hardens first; the interior hardens only to the extent that carbon dioxide in solution or in gaseous form can reach it by penetrating the pores of the material.

SCIENCE AND ENGINEERING

Before the middle of the 18th century, practical knowledge about mortar was passed on from master to apprentice and diffused from one country to another by migrating workmen. The mortar that they made, however, often varied in quality. This variation was not the fault of traditional methods, which were generally sound. The process of manufacturing and using lime consisted of several separate operations over which there was no unified control. Failure to maintain high standards of quality for any one operation could lower the quality of mortar but the inferiority of the mortar might not become known until several years after it had been used in construction. It was difficult to determine what had gone wrong in the mortar-producing process. Any one of the following factors might affect the quality of the mortar: composition of the limerock, lime burning, handling of the quicklime, slaking, composition of the sand, purity of the water, proportioning of lime to sand, thoroughness of mixing the mortar and weather conditions during construction.

In the late 18th century, chemists and engineers began to apply scientific methods to the study of lime and mortar. As the materials became better understood, it was possible to produce mortar of more uniform quality and to calculate its strength with greater accuracy.

NOTED EXPERIMENTERS

Joseph Black (1728–99), a Scottish scientist, was the first to explain why lime was obtained by calcining and why lime mortar hardened. He demonstrated that chalk, when calcined, gave off a gas (carbon dioxide) that he called "fixed air" and that the quicklime remaining after calcination could be reconverted into chalk by exposure to the air. In 1756, he published *Experiments upon Magnesia alba, Quicklime, and some other Alcaline substances*. His experiments also proved that lime obtained from chalk was identical to that obtained from limestone. This discovery helped to discredit the old belief that harder mortar resulted from lime obtained from harder stones. Black, who was professor of chemistry at Glasgow University, was also noted for his quantitative demonstration of latent heat. Knowledge of this demonstration enabled James Watt to greatly improve the steam engine.

John Smeaton (1724–92), an English civil engineer, analyzed lime obtained from various sources and experimented with mortar made from them. By 1756 he proved that limestone containing an appreciable amount of clay could be used in making mortar that would set and harden in water. Smeaton's most famous work was the Eddystone Lighthouse (1759). Many of his experiments and methods were described in the *Narrative of the Building of the Eddystone Lighthouse*, published in 1791.

Bryan Higgins (c. 1737–1820) was a medical doctor who lived in England. He carried out a series of tests on "divers mixtures of lime, sand, and water" and an investigation of "the principles on which the induration and strength of calcareous cements depend" between 1774 and 1780. His *Experiments and Observations*, published by T. Cadell in London in 1780, helped establish a scientific basis for improved practice in preparing and handling lime in building construction. Higgins proved that chalk-lime made mortar "equal, if not superior" to that made with stone-lime. He also demonstrated that, contrary to a widespread custom, fresh lime made better mortar than lime that was kept in paste form for a long time.

> ...*whether Mortar be the better for being long kept before it is used.*
>
> I am generally disposed to think that there is some good reason for any practice which is common to all men of the same trade, although it may not be easily reconcileable to the notions of others: and seeing that the builders slake a great quantity of lime at once, more than they can use for some days, and that all those whom I conversed with, esteemed mortar to be the better for being long made before it is used; and that plasterers particularly follow this opinion in making their finer mortar or stucco for plastering within-doors; I was desirous to discover the grounds of these measures, so repugnant to the notions gathered from the foregoing experiments, and others.
>
> In the month of March 1777, I made about a peck of mortar, with one part of the freshest and best chalk-lime slaked, six parts of sand, and water q. f.; for in a great number of experiments, I observed that this proportion of lime was better than any larger which I had tried, or which the workmen observe in making mortar.
>
> I formed the mortar into a hemispherical heap on the paved floor of a damp cellar, where it remained untouched twenty-four days. At the expiration of this time I found it hardened at the surface; but moist, and rather friable or short than plastic in the interior parts of it.
>
> I beat the whole of it with a little water to its former consistence; and with this mortar and clean new bricks I built a wall eighteen inches square, and half a brick in thickness, in a workman-like manner. On the same day I made a mortar of the same kind, and quantities of fresh chalk-lime and sand, tempered in the same manner; and I built a wall with it, like the former, near it, and exposed equally to the weather.
>
> I examined the mortar in the joints of these walls every fortnight, by picking it with a pointed knife, and could perceive a very considerable difference in the hardness of them; the mortar which was used fresh being invariably the hardest.
>
> At the expiration of twelve months, in pulling these walls to pieces, and by several trials of the force necessary to break the cement and separate the bricks, I found the mortar which had been used quite fresh to be harder, and to resist fracture and the separation of it from the bricks, in a much greater degree than the other specimen....
>
> I concluded that mortar grows worse every hour that it is kept before it is used in building, and that we may reckon as another cause of the badness of common mortar, that the workmen make too much at once, and falsely imagine that it is not the worse, but better, for being kept some time.[19]

Other important authorities on mortar were Louis J. Vicat (1786–1861), chief engineer of roads and bridges in France, and Quincy Adams Gillmore (1825–88), a brigadier general in the United States Army. General Gillmore directed the construction of many fortifications. His *Practical Treatise on Limes, Hydraulic Cements, and Mortars* is worthy of study by those interested in the history of building materials. It was published in 1863 and revised in several later editions.

HYDRAULIC MORTARS

Mortar is said to be *hydraulic* if it will set and harden in water. Hydraulic mortars are of several different compositions:

(1) Common lime plus *pozzolana* (or *trass*, or *arènes*) plus water (sometimes sand)

(2) *Hydraulic lime* plus sand (sometimes omitted) plus water

(3) *Cement* (natural or portland) plus sand (sometimes omitted) plus water (sometimes common lime)

The ingredients in italics impart the hydraulic properties to the mortar.

Pozzolana. *Pozzolana* (often spelled pozzuolana) is a volcanic material sometimes found in a grainy or powdery form and sometimes in gravel-sized pieces or large chunks. It is gray, brown, red or yellow.

The ancient Romans considered pozzolana a kind of sand; they pulverized it (if necessary) and mixed it with common lime to make mortar. The durability of Roman masonry construction gave a high reputation to their materials and methods and led to their emulation by builders in Europe and America.

Pozzolana is inert until mixed with lime and, though bulky, is an easy material to ship and handle without risk of deterioration in quality. When John Smeaton built the Eddystone Lighthouse on the south coast of England in 1759, he used mortar made of slaked Aberthaw lime and pozzolana in equal proportions by volume. Some pozzolana was brought to the United States in the 19th century. The principal source during the 18th and 19th centuries was Civitavecchia, Italy. Other Italian deposits were located near Mount Vesuvius, in the vicinity of Rome and in Sicily. French sources included the Puy-de-Dôme, upper Vienne, upper Loire, Cantal and Vivarais. Pozzolana was also found in the West Indies on the islands of Guadeloupe and Martinique.

Trass. *Trass* (*terras* or *terrass*) is similar to pozzolana. It was found in rock form in large quantities along the Rhine River. Brohl and Andernach were the chief centers for quarrying. At Dordrecht, trass was processed by grinding. Trass was sometimes adulterated by the addition of "wild trass," an inferior material of similar appearance.[20] Trass was widely used in Holland because of its hydraulic properties. Deposits of the material are found on the island of St. Eustasis in the West Indies. Forty tons of trass were purchased in 1796 for the construction of locks on the Middlesex Canal in Massachusetts. At that time, builders in the United States were relatively unfamiliar with the use of trass and it was necessary to experiment with different mixtures in order to determine the proper amounts of trass and lime.

Colonel Loammi Baldwin, engineer for the construction of the canal, used the proportions of two bushels of trass to one bushel of lime and three bushels of sand in mixing mortar.[21]

Dutch Masonry. The Dutch were the leading builders of underwater masonry. Their methods were studied by American engineers working on canal and harbor projects during the 18th and 19th centuries. A description of the Dutch use of trass mortar was printed in *Treatise on Internal Navigation* in 1817:

> They take of the quick-lime about the quantity which will be wanted during a week, and spread it in a kind of bason in a stratum of a foot thick and sprinkle it with water. It is then covered with a stratum of about the same thickness of tarras, and the whole suffered to remain for two or three days, after which it is very well mixed and beaten, and formed into a mass, which is again left for about two days; it is then taken in small quantities, as it is wanted for daily consumption, which are again beaten previous to using. . . .
>
> Tarras is frequently used in this country, being imported from Holland for this purpose. The proportions of the materials of the tarras mortar generally used in the construction of the best water works is the same as the Dutch practice. One measure of quick-lime, or two measures of slaked lime in dry powder, is mixed with one measure of tarras, and both very well beat together, to the consistence of a paste, mixing as little water as possible. Another kind, and almost equally good, and considerably cheaper, is made of two measures of slaked lime, one of tarras, and three of coarse sand; it requires to be beaten a longer time than the foregoing, and produces three measures and a half of excellent mortar. . . .[22]

A similar formula for pozzolana calls for two bushels of slaked Aberthaw lime to one of pozzolana and three of sand.[23] For the finest quality work, sand was not used; where maximum strength was not required, sand was added.

HYDRAULIC LIMES

Hydraulic lime results when "impure" limestone, containing clay in rather finely divided laminations, is burned in the same manner as common lime. Hydraulic lime takes up less water in slaking than pure lime and gives off less heat. It produces a smaller volume of lime paste and

TABLE 9 — COMPOSITION AND PROPERTIES OF HYDRAULIC LIME AND NATURAL CEMENT

Composition of limerock		*Classification*	*Properties*
Lime	*Clay*		
90%	10%	(weakly) hydraulic lime	Will slake when properly calcined, like pure lime, and will harden alone in water.
80%	20%	hydraulic lime	
70%	30%		
60%	40%	natural cement	Will not slake with any degree of calcination and will harden alone in water.
50%	50%		
40%	60%		

generally requires less sand in the mortar. Early American builders probably used a reat deal of weakly hydraulic lime without being aware of its identity. French engineers used hydraulic limes extensively in civil construction during the 19th century. Several kinds of English lime are hydraulic, notably Lias lime, which is still used for repair and restoration work. Aberthaw lime is weakly hydraulic.

The distinction between hydraulic lime and natural cement is arbitrary. Hydraulic lime can generally be slaked but natural cement cannot. The classifications in Table 9[24] are adapted from those of Petot, a French civil engineer.

TABLE 10 — MANUFACTURE OF NATURAL CEMENT IN AMERICA

Date	*Location*
1824	Williamsville, Erie County, N.Y.
1826	Kensington, Conn.
1828	Rosendale, Ulster County, N. Y.
1829	Louisville, Ky.
1831	Williamsport, Pa.
1836	Cumberland, Md.
1837	Round Top, near Hancock, Md.
1838	Utica, Ill.
1839	Akron, N. Y.
1848	Balcony Falls, Va.
1850	Siegfried's Bridge, Lehigh Valley, Pa.
1850	Cement, Ga.

NATURAL CEMENT

The terms *hydraulic cement, waterproof cement* and *water-lime* were commonly used in the United States before about 1870; since then they have been supplanted by the term *natural cement.* Natural *cement rock* was burned in kilns similar to those used for lime; the calcined lumps were then ground into a fine powder and the cement was stored in airtight waterproof containers. Refinements in this basic process were introduced late in the 19th century.

Early Manufacture. John Smeaton came very close to making natural cement but the first person to patent the process was another Englishman, James Parker, whose patent was granted in 1796, 40 years after Smeaton's experiments. *Parker's Roman Cement,* which was advertised as equaling that used by the ancient Romans, was made from stones called *nodules* or *septaria* found at Harwich and Sheppy. Parker's product enjoyed a good reputation in Europe and America for several decades. He recommended making mortar by mixing two measures of water with five of Roman cement. This mortar set quickly (in 10 to 20 minutes), a quality that was considered a disadvantage. After Parker's patent expired, Roman cement was widely manufactured by others, who used cement rock obtained at several places in England and Boulogne pebbles found at Boulogne, France.

Edgar Dobbs of London obtained a patent in 1810 for "artificial hydraulic lime," which is better described as artificial cement. Calcium carbonate and clay were mixed in a wet state, dried, cut into pieces and calcined. The resulting product was pulverized or ground.

Natural cement rock was first discovered in the United States, somewhere between Sullivan and Fayetteville, N.Y., by Canvass White (1790–1834) and other engineers directing construction on the Erie Canal. In 1820, White obtained a patent on his method of manufacturing natural cement, and about 1825 he and his brother, Hugh, established a factory at Chittenango, N.Y. They called their product White's Water-Proof Cement. Hugh White later manufactured natural cement at Whiteport, in Ulster County, N.Y.[25]

Natural cement rock was soon discovered in other localities, and the widespread manufacture of natural cement was begun. Table 10[26] gives a listing of manufacturing locations.

New York was for decades the country's largest producer of natural cement. Cement was shipped down the Hudson River to ports along the Atlantic coast and to the West Indies.

Natural Cement Mortar. After 1819, all masonry used in the construction of the Erie Canal was laid in natural cement mortar. Various sources afford different information about the mortar mix; apparently one part of sand was mixed with two parts of cement. The general practice in New York State in about 1840 was to mix two or three parts of sand to one of cement. According to H.S. Dexter, a civil engineer reporting to the New York state legislature at that time, those amounts of sand were excessive.[27]

Natural cement mortar was used mainly in areas where masonry was subjected to moisture and great strength was required. It was rarely used except for buildings of the finest construction. Natural cement shrank in volume when water was added; masons did not like the "feel" of cement mortar as well as that of lime; and the color of cement mortar was sometimes unpleasant in appearance. However, natural cement was sometimes used as an additive to lime mortar, to which it contributed strength and durability. Cement mortar was also used by the federal government in building forts, engineering works and public buildings, including the extension of the Capitol in Washington, D.C., in the 1850's. The architect, Thomas U. Walter, wrote: "I have to request that all the mortar used in every part of the work be mixed in proportions by actual measure, as follows. For all the *footings*, to the height of 2 feet above the bottom of the cellars, and for the backing behind the granite subbasement, *cement* and *sand*, without any lime, in proportions of *one* of cement to *two* of sand. For all the rest of the work, in proportions of one of cement, three of lime and eight of sand."[28]

PORTLAND CEMENT

Joseph Aspdin (1779–1855) was an English mason-builder who was interested in producing artificial stone. In 1824 he patented a material called Portland Cement. By the middle of the century, improved portland cement was widely used in England. It was manufactured in the United States from 1871 on and gradually replaced natural cement for many purposes, partly because it was more dependably uniform in quality. By the late 19th century, the demand was such that portland cement was imported into the United States from England, Germany, Belgium and other European countries.

Portland cement became a major ingredient in mortar after 1880. Its strength, low absorbency and hardness were well matched to the bricks made during that period. In 1896, F. E. Kidder gave the following recommendations for mortar in his *Building Construction and Superintendence:*

> Cement mortar should be used for all mason work which is below grade, or situated in damp places, and also for heavily loaded piers and arches of large span. . . .
>
> For construction under water, and in heavy stone piers or arches . . . Portland cement should be used; elsewhere material cement will answer.
>
> For natural cements the proportion of sand to cement by measurement should not exceed 3 to 1, and for piers and first-class work 2 to 1 should be used. Portland cement mortar may contain 4 parts of sand to 1 of cement for ordinary mortar, and 3 to 1 for first-class mortar. For work under water not more than 2 parts of sand to 1 of cement should be used. When cheaper mortars than these are desired it will be better to add lime to the mortar instead of more sand.
>
> In making cement-lime mortar the sand and cement are thoroughly mixed dry, the lime putty is mixed with water and screened into a mortar box, and the whole is then thoroughly mixed and worked together until a proper consistency is obtained.
>
> The following are mixtures by measure that have been used with excellent results:
>
> Cement 1 part, sand 8 parts, lime paste 1½ parts.
> Cement 1 part, sand 6 to 7 parts, lime paste, 1 part.
> Cement 1 part, sand 8 parts, lime paste ½ parts.
> Cement 1 part, sand 10 parts, lime paste 2 parts.[29]

MORTAR COLOR

The clean, white appearance of lime tinted slightly by sand was a highly favored architectural effect. White marble dust was sometimes added to mortar, replacing part or all of the sand, when pointing the joints between bricks and stones. Colored mortar, obtained by mixing in mineral or earth pigment, has been used sparingly since ancient times. Common lampblack and Venetian red were among the pigments employed; they were subject to fading. During the second half of the 19th century, dark mortar was popular. The narrow joints then fashionable, when colored to approximate brick color, contributed to an effect of continuity of the wall surface. Cement mortar had a gray color which lent itself well to dark joints; however, when light mortar was required, cement mortar became undesirable. Eventually, special white cement was developed for use in light-colored mortar.

As a general rule, the color of the mortar used in historic buildings in the United States depended on the color of the sand used in the mixture.

MASONRY JOINTS

The pattern and detailing of mortar joints on the face of a brick or stone wall greatly affects their appearance and resistance to weathering. Joints vary in thickness for both structural and aesthetic reasons. When a wall is laid with irregular masonry units, joints must be wide enough to accommodate variations in size and shape. Bricks or stones of uniform size can be laid with joints as thin as the properties of the mortar permit. The minimum required thickness for masonry joints is about three-sixteenths of an inch in order for mortar to constitute a coherent material that will adhere to porous substances. A thicker cushion of mortar is often desirable. A joint up to three-eighths of an inch thick or more is usual for brickwork;joints in stone rubble work are often much thicker. Mortar in thick joints may shrink appreciably in setting.

Pointing. The exposed parts of joints were sometimes filled with mortar after that used in the interior had hardened. This process, called *pointing*, was done when the building was approaching completion. Sometimes a more durable mortar than that used on the interior was used for pointing. The purposes of pointing were to secure a particular appearance or to make the exterior of the wall more durable.

Monumental buildings of ashlar were usually pointed, especially after cement mortars came into use. However, cement mortars stain granular limestone, marble and some kinds of sandstone. To avoid stains, non-staining cements of lime, plaster of paris and marble dust were developed.

Brick Joints. Different treatments are used in finishing brick joints. The simplest is the scraping off of excess mortar from the face of the wall with the mason's trowel, leaving a *flush joint*. A *struck joint* is made by holding a trowel obliquely and drawing the point along the mortar. In exterior work, the face of the struck joint should slope inward from the top, forming a *drip*. A *jointer* or *jointing tool* of the proper width to fit into a joint is used to press the exterior mortar into a concave profile. Such *tooled joints* were often *ruled* with the aid of a straightedge; a narrow tool was used to inscribe a line along the center of the joint. *Raised joints* are made with a trowel and straightedge. End joints and vertical joints were sometimes thinner than horizontal ones; when bricks varied in length the differences were taken up in the joints.

Stone Joints. The irregular pattern and thickness of joints in rubble called for varied treatment with the trowel. Sometimes stones projected irregularly beyond the face of a flat surface of mortar. Often a V-shaped ridge profile was given to the mortar, while at other times a flat raised joint was preferred. Occasionally, pebbles were pushed into the mortar before it set, giving a decorative effect known as *galleting*.

Ashlar joints are similar to brick joints. Ashlar work on monumental buildings is frequently *rusticated;* the stones are cut so as to form flat or beveled channels between adjacent pieces. When this is done, mortar joints are subordinate to the channels. Drafted margins on rock-faced stones produce a comparable effect; the actual mortar joint becomes an inconspicuous element.

DETERIORATION OF MORTAR

Deterioration brought about by moisture in walls, described in the chapters on

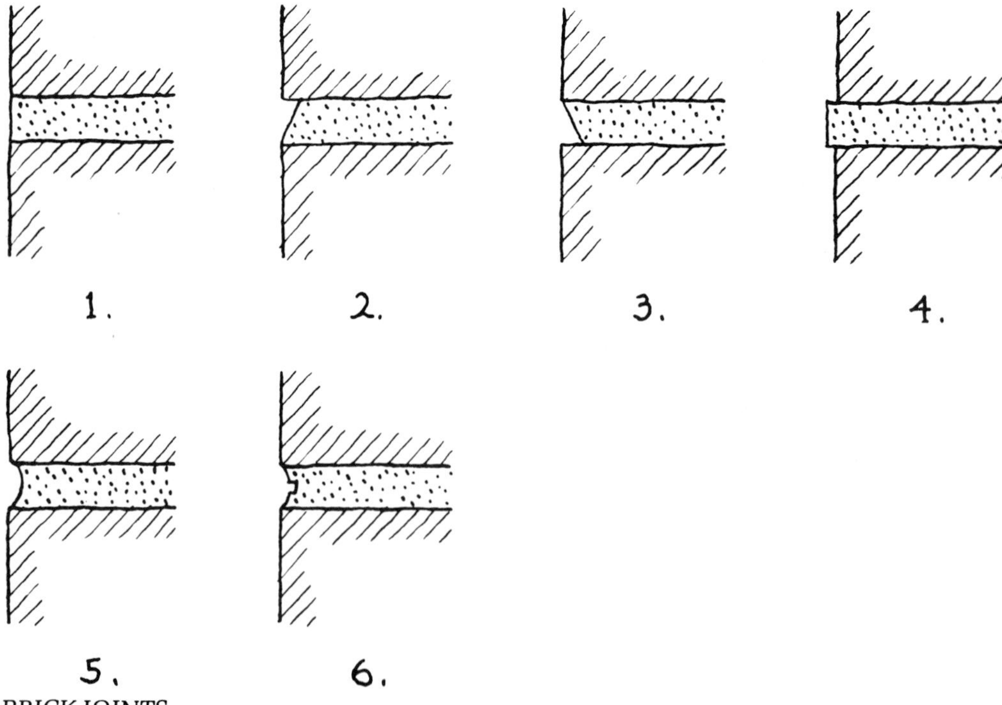

BRICK JOINTS.
1. Flush. Surplus mortar was scraped off with a trowel.
2. Struck, with drip. Done with the point of the trowel.
3. Struck, without drip. Not for exterior exposure.
4. Raised.
5. Tooled. Done with a jointer.
6. Tooled and scribed (or ruled).

stone and brick, also takes place in mortar. The mortar inside an old wall is a material of unknown quality. If, when the building was constructed, the interior mortar simply dried out instead of hardening and never adhered to the bricks or stones, the mortar was little more than a dry mixture of sand and lime from which the lime could readily leach out.

In old houses whose chimneys were constructed without flue linings or were lined with plaster, lime mortar was not greatly affected by gases (and condensation) as long as wood was burned in the fireplaces. When anthracite coal came to be used as a fuel, however, the mortar was seriously damaged.

REPOINTING

Repointing is probably the most common operation practiced in preserving and restoring old masonry buildings. If well-done, it is not only a safeguard to the physical structure, but an important contribution to the maintenance of the historical character of the wall. Improper repointing with soft mortars was sometimes done in the past, but repointing that has been done since the introduction of hard cement mortar is more harmful. Repointing when badly done is difficult and expensive to correct; in extreme cases it causes irreparable damage to the physical structure of the building as well as its appearance.

Galleting in foundation joints. John Ridout House (c. 1760), Annapolis, Md.

Repointed raised joints in a regular pattern between irregular gneiss stones. Daniel MacKinett House (by 1759), Germantown, Philadelphia, Pa.

Raised joints in an irregular pattern on brownstone. Recently repointed with hard mortar, a number of stones have deteriorated by exfoliation, especially those to the left of and above the tablet. Barn or mill (1785), Lewisberry, Pa.

Before repointing, all loose and deteriorated mortar should be removed. The joints must be raked out by hand to a depth of about one inch; care must be taken to avoid enlarging the width of the joints.

Choice of Mortar. When choosing the type of mortar to be used in repointing, full consideration must be given to matching the old mortar in color, texture, strength and hardness (density and porosity). The color can be approximated by using materials similar to those of the original mortar. The original mortar can be analyzed by means of a simple test. A sample of the old mortar should be crushed and shaken up in a glass of water. Then the appearance of the particles that settle out may be studied. This analysis also helps in identifying ingredients that will produce the desired texture.

Stippling the joint (marking it by touching it with the end of a stiff brush) before the mortar completely sets helps to give it a worn appearance.

The new mortar used in repointing should have the same physical characteristics as the old only if the old mortar was reasonably appropriate in the first place. It is best to repoint with mortar having the same density and absorbency as the stones or bricks in the wall. Soft bricks and stones should be repointed with soft mortar; hard cement mortar will cause the softer materials to disintegrate. T.A. Bailey of the Ancient Monuments Branch of the British Ministry of Public Buildings and Works mentions this effect in *Notes on Repair and Preservation: Masonry, Brickwork.*

> Cement pointing is detrimental, particularly if it is applied to soft stone (or bricks) because it is hard, non-resilient, comparatively non-absorbent and therefore does not respond to the variations in the atmosphere to the same extent as the stone or brickwork with which it is in contact. If hard pointing is employed the physical action causes rapid weathering and disintegration of the softer stone or brickwork. Many cases of stone decay have been directly traceable to a porous stone being pointed with impervious mortar. In such cases both saturation and evaporation are confined to the stone whereas the process should be evenly distributed over stone and pointing.
>
> Soft hand-made or underburnt bricks, deteriorate rapidly when they are set and/or pointed up in Portland or other hard setting cement. In fact, the greater the difference in density of the bricks to the mortar,

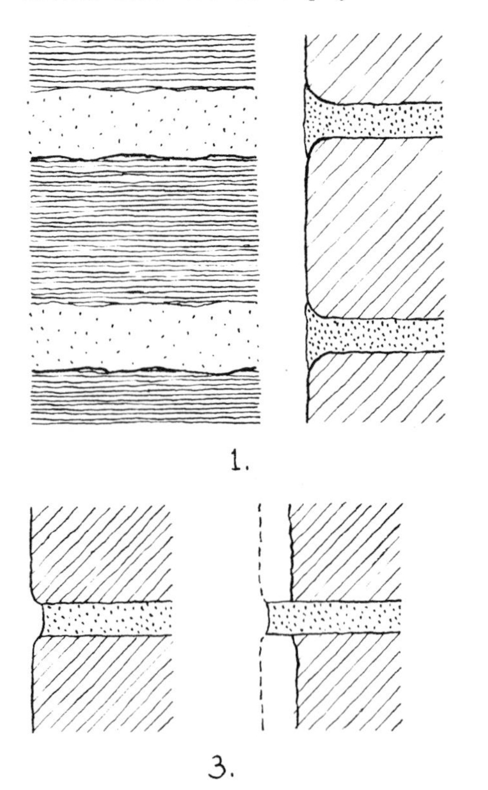

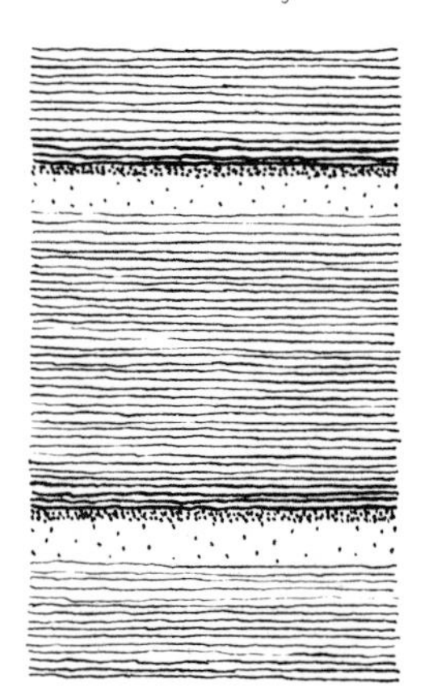

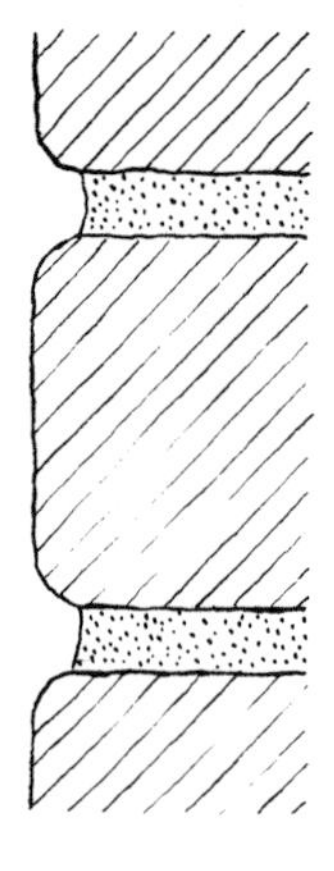

2.

REPOINTING.
1. Old, weathered brick or stone often has worn, rounded edges. Feather edges of mortar break off, taking with them particles of brick. Dirt and water enter the cavities and hasten deterioration.
2. Recessed joints look better and are less susceptible to damage.
3. When soft bricks are pointed with hard cement mortar, the hard mortar remains but the bricks disintegrate.

the greater is the degree and rapidity of disintegrating of the bricks. Examples of this form of erosion are often to be seen where the hard mortar remains but the brickwork surface has receded several inches behind the original face. This erosion is due to the fact that practically all moisture absorbed in the wall is held in suspension in the softer textured bricks and when frost supervenes the surface of the brick gradually disintegrates.[30]

An even more frequent occurrence is the cracking of hard cement mortar when used with bricks and stones for which it is not appropriate. Cracking of cement mortar may also result from faulty application.

For repointing of old soft-brick buildings, preservation architects in the United States often recommend lime-sand mortar mixed with portland cement in a ratio of one part cement to two or three parts of lime. British architects at the Ancient Monuments Branch prefer two parts of hydraulic quicklime to five of sand. For a quicker setting mixture, they use some portland cement, not in excess of one part to 12 of the lime-sand mixture.

Repointing Weathered Materials. Weathered bricks and stones in an old wall frequently acquire worn edges and rounded profiles. When repointing them, it is advisable to recess the face of the new mortar slightly to keep the joint from becoming too wide and to avoid spreading mortar over the edges of the bricks and stones. When repointing bricks and rubble, feather edges should be avoided; they break off easily, carrying particles of stone with them and leaving cavities through which moisture may enter.

TUMBLING

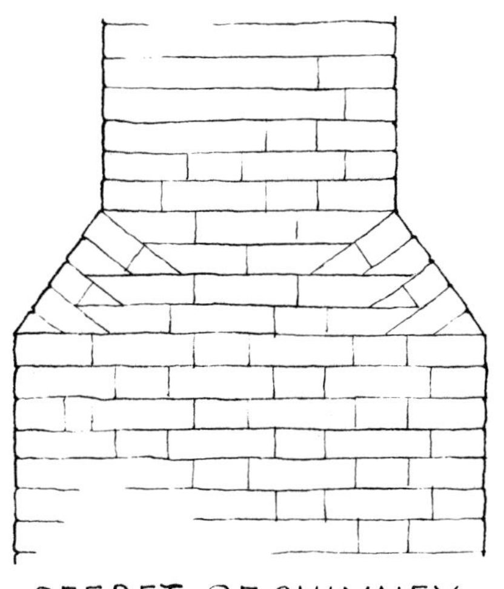

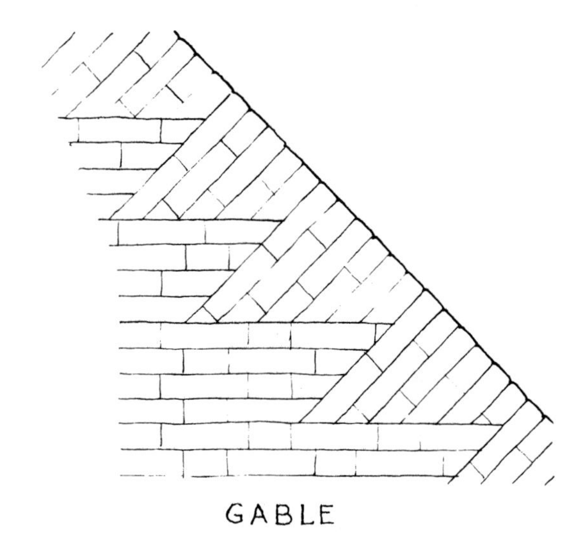

RACKING TOOTHING

TUMBLING. This manner of laying was sometimes used to finish the sloping edges of a brick wall. Tumbling was most common in colonies settled by the Dutch.

RACKING AND TOOTHING. When one part of a brick wall is temporarily built up higher than the rest, the end of the higher part may be stepped or racked back to the connect with the lower level. Toothing is inferior to racking.

Conglomerate rubble with cut brownstone trim. St. Thomas Episcopal Church (1846), Glassboro, N.J. John Notman, architect. (Jack E. Boucher for Historic American Buildings Survey.)

Horizontal rustication, unfluted monolithic columns, random ashlar wall with dressed trim around entrance. South entrance porch, Chester County Prison (1838), West Chester, Pa. (Ned Goode for Historic American Buildings Survey.)

Repointed wall. The mortar is slightly recessed and does not cover the bricks' edges. Joseph and William Shippen House (c. 1744), Philadelphia, Pa.

Repointed sandstone. This work may or may not resemble the original mortar joints but is sympathetic in appearance to the old wall. 129 Court Street, Newtown, Pa.

Nineteenth-century repointing with colored mortar. Lines have been painted over joints. Philadelphia, Pa.

Careless repointing, especially on the right. Philadelphia, Pa.

Eighteenth-century split lath. First-floor ceiling, Philipse Manor, Yonkers, N.Y. (John G. Waite for the Division for Historic Preservation, New York State Office of Parks and Recreation).

Deterioration of sandstone pointed with hard mortar. The surface of the stones is crumbling off; several have lost as much as one inch of material. Newtown, Pa.

Shallow repointing of a mid-18th-century limestone wall. The hard mortar has cracked and fallen off in many places. A number of stones are deteriorating by exfoliation. Easton, Pa.

Painted lines give mortar joints the appearance of regularity. Belle Grove (1794), Middletown, Va. (William Edmund Barrett)

FOOTNOTES FOR CHAPTER III

1. Arthur W. Brayley, *History of the Granite Industry of New England*, vol. 1 (Boston: National Association of Granite Industries of the United States, 1913), p. 10.
2. J. Frederick Kelly, *The Early Domestic Architecture of Connecticut* (New Haven, Conn.: Yale University Press, 1924), p. 70; Rosalie Fellows Bailey, *Pre-Revolutionary Dutch Houses and Families in Northern New Jersey and Southern New York* (New York: William Morrow & Company, 1936), p. 23.
3. *Bulletin of The Association for Preservation Technology*, vol. V, no. 1 (1973), p. 71. Specification transcribed by Hugh C. Miller.
4. Observations made by the author in 1959.
5. Quoted by Marcus Whiffen, *The Eighteenth-Century Houses of Williamsburg* (Williamsburg, Va.: Colonial Williamsburg, 1960), pp. 7–8.
6. Kelly, *op. cit.*
7. Observations made by the author in 1962–63.
8. Joshua V. H. Clark, *Onondaga*, vol. 2 (Syracuse, N. Y.: Stoddard & Babcock, 1849), p. 67.
9. Sylvanus Griswold Morley, *The Ancient Maya*, 2nd ed. (Stanford, Calif.: Stanford University Press, 1947), pp. 350–51.
10. Measured drawings of several lime kilns in eastern Pennsylvania are reproduced in the *Proceedings of the Lehigh County Historical Society*, volume 19 (August 1952).
11. Norman Davey, *A History of Building Materials* (London: Phoenix House, 1961), p. 101.
12. Col. Joseph G. Totten, *Essays on Hydraulic and Common Mortars and on Lime-burning*. Translated from the French of Gen. Treussart, M. Petot and M. Courtois. With Brief Observations on Common Mortars, Hydraulic Mortars, and Concretes, and an account of some experiments made therewith at Fort Adams, New Port Harbour, R. I., from 1825 to 1838. (Philadelphia: Franklin Institute, 1838), p. 11.
13. Heinrich Ries and Edwin C. Eckel, *Lime and Cement Industries of New York* (Albany, State of New York: 1901), pp. 675–76.
14. Vitruvius, *The Ten Books on Architecture*, trans. Morris Hicky Morgan (reprint ed., New York: Dover, 1960), bk. 2, chap. 5, p. 1.
15. C. F. Partington, *The Builder's Complete Guide* (London: Sherwood, Gilbert and Piper, 1825), p. 364.
16. Norman M. Isham and Albert F. Brown, *Early Rhode Island Houses* (Providence, R. I.: Preston & Rounds, 1895), p. 73.
17. Rosamond R. Beirne and John H. Scarff, *William Buckland* (Maryland Historical Society, 1958), pp. 24–25.
18. Peter Nicholson, *The New Practical Builder* (London: Thomas Kelly, 1823), p. 354.
19. Bryan Higgins, M. D., *Experiments and Observations* (London: T. Cadell, 1780), pp. 35–38.
20. *Encyclopaedia of Architecture* (Amsterdam and Brussels: Elsevier Publishing Company, 1954). See *Trass*.
21. Christopher Roberts, *The Middlesex Canal 1793–1860* (Cambridge, Mass.: Harvard University Press, 1938), pp. 94–99. A small ship was sent to St. Eustasis expressly to bring this material to Massachusetts.
22. *Treatise on Internal Navigation* (Ballston Spa, N.Y.: U.F. Doubleday, 1817), chap. 14.
23. Peter Nicholson, *The Mechanic's Companion* (Philadelphia: James Locken, 1832), p. 187.
24. Totten, op. cit., p. 191.
25. Harley J. McKee, "Canvass White and Natural Cement," *Journal of the Society of Architectural Historians*, vol. 20, no. 4, pp. 194-97. This product should not be confused with a late 19th-century English product called White's Cement.
26. Compiled after data in Uriah Cummings, *American Cements* (Boston: Rogers & Manson, 1898).
27. *State of New-York, No. 123, In Senate, May 6, 1840* (Albany: State of New York, 1840), pp. 16–17.
28. Thomas U. Walter to Samuel Strong, 14 April 1852, Office of the Supervising Architect of the United States Capitol, Washington, D.C. This reference is through the courtesy of Charles E. Peterson, FAIA, FRSA.
29. F. E. Kidder, *Building Construction and Superintendence*, 9th ed., rev. by Thomas Nolan (New York: Comstock, 1910), pp. 188–89.
30. T. A. Bailey, "Notes on Repair and Preservation: Masonry, Brickwork," mimeographed (London: Ministry of Public Buildings and Works, 1960), Masonry, p. 1; Brickwork, p. 1.

IV. PLASTER

INTRODUCTION

Plaster, an important element in historic buildings of all periods, is used to cover (exterior and interior) walls and ceilings. It is applied in a plastic condition and hardens into a uniform, smooth finish. Not only does it afford a pleasing surface that can be ornamented if desired, but it helps to make structures more resistant to penetration by wind and rain. The components of plaster are basically the same as those of mortar: clay, lime, gypsum and sand. The term *mortar* is frequently applied to the mixture used in ordinary plastering.

The word *pargetting* (pergetting, pergening, parging, parge-work) was in use as early as 1450 to denote a plastic covering for walls and ceilings. The term more commonly referred to ornamental work but it was also applied to plain work. *Plaster* (plaister) is another early word for such "daubed-on" coverings; it was applied to both interior and exterior work. The word *stucco* was used to denote ornamental plasterwork and finely finished material. Since the 19th century, the term stucco has been favored in the United States when referring to plaster on the exterior of walls.

CLAY PLASTER

Clay or mud plaster was in use at least as early as 3500 B.C. in Mesopotamia; it was probably employed in buildings much earlier. Settlers in the American colonies used clay plaster on the interiors of their houses: "Interior plastering in the form of clay antedated even the building of houses of frame, and must have been visible in the inside of wattle filling in those earliest frame houses in which ...wainscot had not been indulged. Clay continued in use long after the adoption of laths and brick filling for the frame. Records of the New Haven colony in 1641 mention clay and hay as well as lime and hair.... In the German houses of Pennsylvania the use of clay persisted much later still."[1]

A contract (c. 1675) from Salem, Mass., specifies the use of clay plaster:

> ... he is to lath and siele the 4 rooms of the house betwixt the joists overhead with a coat of lime and haire upon the clay; also to fill the gable ends of the house with bricks and plaister them with clay. 5. To lath and plaister partitions of the house with clay and lime, and to fill, lath and plaister them with lime and hair besides; and to siele and lath them overhead with lime; also to fill lath and plaister the kitchen up to the wall plate on every side. 6. The said Daniel Andrews is to find lime, bricks, clay, stone, haire, together with laborers and workmen. . . .[2]

In New Mexico and other states of the Southwest, clay was used to plaster adobe walls:

> The walls are plastered with the same earth that was used for the bricks. Yet nature provides an astonishing variety of soft colors that make for extremely beautiful interior "plaster." Not every soil, however, is appropriate for plastering. Usually each community has a clay pit where a usable mud can be obtained for plaster, a fact which accounts for the uniform color of the local houses. The earth selected is carefully screened and applied with bare hands. When the plastered area has dried, it is smoothed over once more with a piece of dampened sheepskin. . . . A dado of darker colored adobe plaster was often used around the lower part of the wall. . . When a lighter interior was desired, a coat of calcimine was brushed on over the mud plaster.[3]

LIME PLASTER

A mixture of lime and sand with hair or other substances was used for plastering since ancient times. It was applied directly on masonry or on a base of wood lath. Archaeologists who investigated the remains of buildings important to early Western civilization have discovered lime in the plasterwork of walls, ceilings, floors and roof coverings. Samples from Crete dated c. 2100 B. C. contained about 40 percent lime; other ingredients were silica (sand), alumina (clay), fragments of pottery and chopped straw. The plaster was about five centimeters thick, the outer coat being finer in texture than those underneath and accounting for one and a half centimeters of the total thickness.[4] Lime plaster was used for floors and wall coverings at Zaculeu, Guatemala, by c. 300 A.D.: "Hard white plaster covered the stonework to a thickness of 2 cm. It was of fair quality and had been applied as a very wet mixture, filling the interstices of the masonry, but the surface was unevenly finished and showed no signs of smoothing or polishing. . . . The floor averaged 4 cm. in thickness. . . A well-proportioned amount of coarse aggregate gave it a concrete-like quality and the surface was evenly finished."[5]

Lime was the principal ingredient used in plaster in the United States. Lime plaster was in common usage in New Haven, Conn., by 1641, and by 1700 it was being used in Hartford, Wethersfield and Windsor, as well.

> Examination of many examples of early plastering reveals the fact that it is generally "one-coat work,"

and that, although rough in texture and finish, it is of great hardness and durability. Shell lime seems often to have entered into its make-up, especially in towns along the Sound, as well as a generous amount of red cattle hair. . . . Such specimens of early plaster work are always very rich in lime; and where the source of it was oyster shells, it is common to find good-sized fragments of them, imperfectly calcined, in the plaster.[6]

Lime and sand plastering was the usual wall covering inside inexpensive cottages in 1850; it as often whitewashed.[7]

To prepare plaster for the first coats, quicklime was slaked in a basin or vat in an excess of water for several days. Thorough slaking was essential, because if any unslaked lumps remained in the plaster after it was applied to the wall, they would eventually slake and expand, causing pieces of plaster to pop off. The plaster or mortar used in the first coats was called *coarse stuff.* To make coarse stuff, about one-sixth part (by volume) hair was mixed with one part lime paste and two to two and one-quarter parts sand were added. Sand with rounded grains was preferred.

For finish coats, *fine stuff* was prepared. Lumps of quicklime were slaked to a paste with a moderate volume of water and then diluted to the consistency of cream. This mixture was allowed to harden by evaporation to the desired consistency for working. Fine white sand, marble dust or plaster of paris were sometimes added. About one part of sand was mixed with three parts of lime paste. Marble dust was generally reserved for ornamental plasterwork. When plaster of paris was added, the material was called *gauge stuff; gauging,* or adding plaster of paris, made the plaster set more quickly.

GYPSUM PLASTER

Gypsum is a rock found in many parts of the world. It has been used to make walls and pavements in places sheltered from rain but its primary use is in making plaster for walls, ceilings, ornamental work and even for floor coverings. During the Middle Ages, gypsum quarries in the vicinity of Paris supplied large amounts of the material to make plaster, giving rise to the name plaster of paris. That term is used today to refer to calcined gypsum from any source.

The chief source of gypsum in North American during colonial times was Nova Scotia; some was imported into the eastern colonies. Extensive deposits were discovered in central New York near Camillus in about 1790. Quarries worked there after 1792 supplied much gypsum for fertilizer as well as for plaster. Numerous mills for grinding plaster, run by waterpower, were built along Chit-

Exterior clay plastering on wood lath. A broken place was photographed to reveal the wall construction; clay wall coverings can be astonishingly durable when protected by wide overhanging roof. Kashiwara-Shi, Japan.

tenango Creek and other streams in the region.

Gypsum is dihydrate of calcium sulphate. It is heated in open kilns to a temperature of 270 to 340 degrees F. for about three hours, losing about three-fourths of its water of crystallization, to become a hemihydrate. After cooling, lumps of hemihydrate are ground into a fine powder (plaster of paris). If gypsum is heated to 390 degrees F., it loses all of its water of crystallization. Such material was once considered "dead" but is now used in the production of anhydrous plasters. Keene's cement, patented in 1836, is the best known of these. It is made by adding alum to gypsum and heating the mixture to 1100 degrees F. Since its introduction, Keene's cement has been widely used on the walls of kitchens and bathrooms because of its hardness and resistance to moisture.

PLASTERING

Plasterers learned their trade by apprenticeship. Those who did ornamental work were usually distinguished from those who plastered plain walls and ceilings. Ornamental plaster workers who modeled wet plaster in situ were sometimes called *stuccadors*, a name derived from that of the Italian workmen who introduced that technique into northern Europe.

Tools. A spade and a rake were used for mixing plaster; the latter had either two or three prongs and was employed for mixing in hair. Two kinds of trowels were used to apply and smooth plaster. One was a thin piece of hardened iron that was about 10 inches long and two and a half inches wide. It was rounded at one end. The other kind of trowel was rectangular; it was made in several sizes. Both kinds of trowels had wooden handles attached to the back with a bent piece of iron. Triangular trowels were used for forming moldings. A trowel was sometimes called a *float*. A *long float*, also called a *darby*, was about four feet long and seven inches wide; it was manipulated by two workmen and consequently had two handles. Specially shaped floats were used to finish corners. The plasterer's *hawk* was a flat piece of wood 10 inches square with a handle attached to the bottom A workman held the hawk, on which a small quantity of plaster was placed, in his left hand and worked with the trowel in his right hand. For smoothing the finish coat of plaster, a brush was often used to supplement the trowel.

Operations. Plaster was applied to walls and ceilings in two or three coats. The coats were given different names when applied to lath or to masonry. The final coat was named according to its finish. The terms in Table 11[8] were in common use in England and the United States during the 19th century.

Two-coat plastering on lath could generally result in a plane wall surface if *grounds* were installed to guide the workman (grounds are narrow boards at the base of a wall and at door and window openings). On ceilings, where no grounds

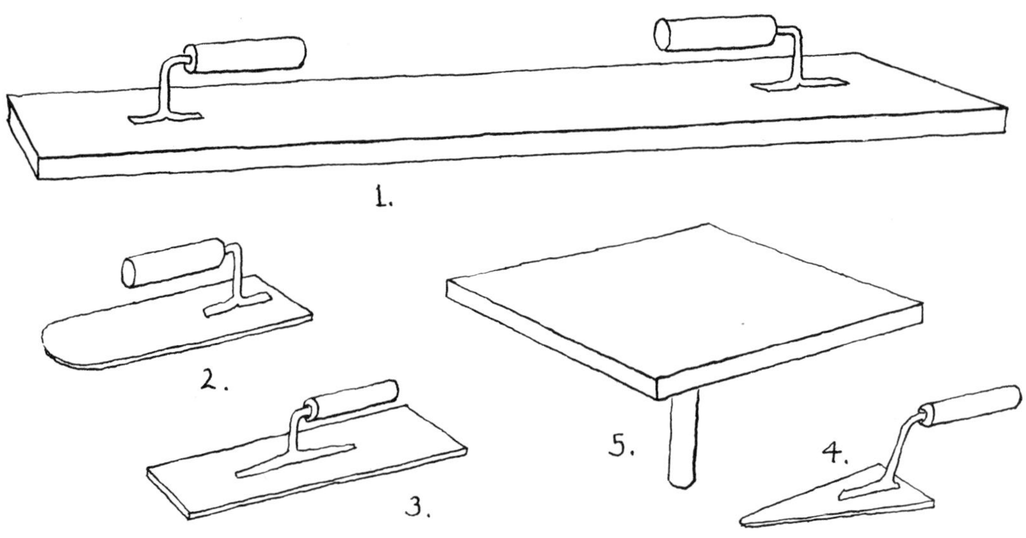

PLASTERER'S TOOLS
1. Long float or darby.
2. Laying and smoothing trowel, 10 inches long.
3. Modern trowel.
4. Trowel for forming cornices and moldings, 2 to 7 inches long.
5. Hawk, 10 inches square.

were installed, it was difficult to avoid a rolling surface. To correct the normal irregularities of masonry walls, three-coat work was required. The second coat was sometimes applied as soon as the first had set sufficiently to receive it, but in better practice the first coat was allowed to dry thoroughly before the second coat was applied. *Screed work* or *screeding* was once the standard for good three-coat plastering, but by the end of the 19th century it was reserved for only the most expensive jobs. In this type of work, strips *of plaster called screeds* that were equal in thickness to the next coat were applied to the wall at intervals of two to four feet. Screeds were from six to eight inches wide. After they dried, the areas between screeds were plastered in.

In floated work these screeds are made, at every three or four feet distance, vertically round a room, and are prepared perfectly straight by applying the straight-edge to them to make them so; and when all the screeds are formed, the parts between them are filled up flush with lime and hair, or *stuff*, and made even with the face of the screeds. The straight-edge is then worked horizontally upon the screeds, to take off all superfluous *stuff*. The floating is thus finished by adding *stuff* continually, and applying the rule upon the screeds till it becomes, in every part, quite even with them. . . .
Ceilings are floated in the same manner, by having screeds formed across them. . . .[9]

Kidder, near the end of the 19th century, spoke of wall screeds as being laid horizontally.[10]

Plain plasterwork was measured and priced by the square yard.

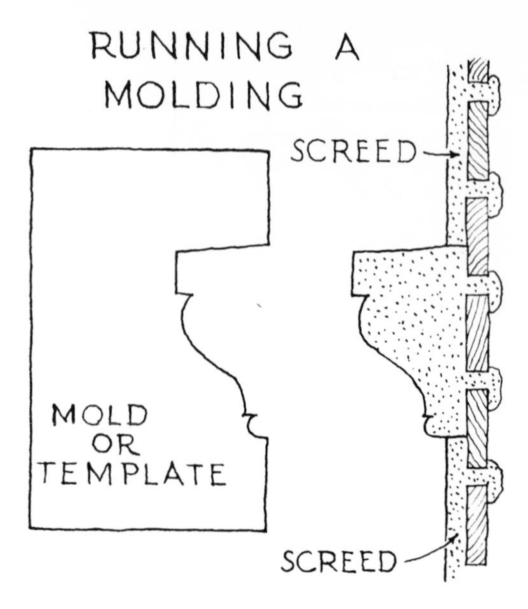

EXTERIOR PLASTER

Exterior walls have been covered with plaster since ancient times, notably in Mediterranean lands. During the early 19th century, exterior plastering was fashionable in England. Several American centers that maintained close cultural ties with England followed this practice. In Charleston, S. C., a thin coating of limestone mortar (only one-eight to one-quarter inch thick) was applied to brick walls, columns and molded trim. The term *rough cast* was applied to this covering in England but in the United States it was usually called *stucco.*

Stucco Mixtures. In the northern states, stucco walls were not numerous. Accord-

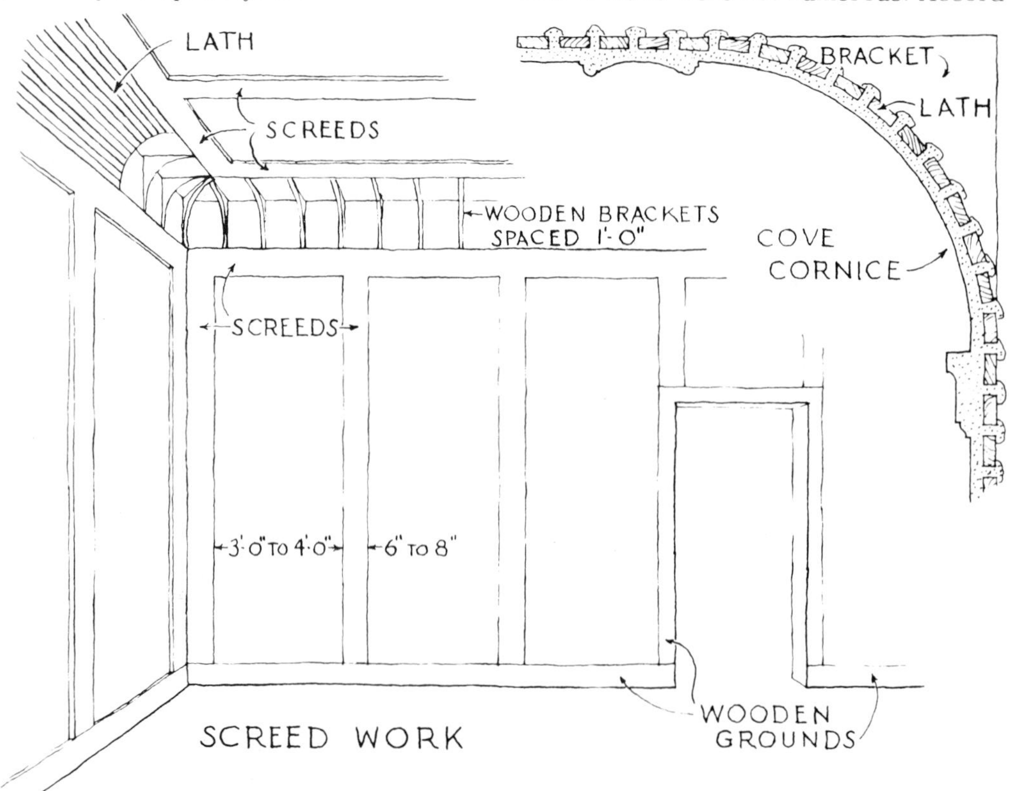

ing to Andrew Jackson Downing in *The Architecture of Country Houses:*

Outside plastering or stuccoing is generally so little understood in this country, and has been so badly practiced by many masons, that there is an unjust prejudice against it, in many parts of the Union .... A strong and durable stucco for plastering the outsides of rough brick or stone walls... Take stone lime fresh from the kiln, and of the *best quality*...slake it by sprinkling...to...a fine *dry powder*, and not a *paste*. Set up a quarter-inch wire screen at an inclined plane, and throw this powder against it. What passes through is fit for use. That which remains behind contains the *core*, which would spoil the stucco, and must be rejected.

Having obtained the sharpest sand to be had, and having washed it so that not a particle of the mud and dirt (which destroy the tenacity of most stuccoes) remains, and screened it, to give some uniformity to the size, mix it with the lime in powder, in the proportion of *two parts sand* to one part lime. This is the best proportion for lime stucco. More lime would make a stronger stucco, but one by no means so hard—and hardness and tenacity are both needed.[11]

Downing cited houses in Pennsylvania on which common lime stucco stood up well for a century. He thought that some hydraulic limes and natural cements were suitable for stucco work and suggested that only those that had been tested for at least 10 years of outdoor exposure should be used.

Roman cement was specified for the exterior work on Beth Elohim Synagogue in Charleston, S. C. This building had fluted Greek Doric columns worked in brick. The overhang of the cornice was

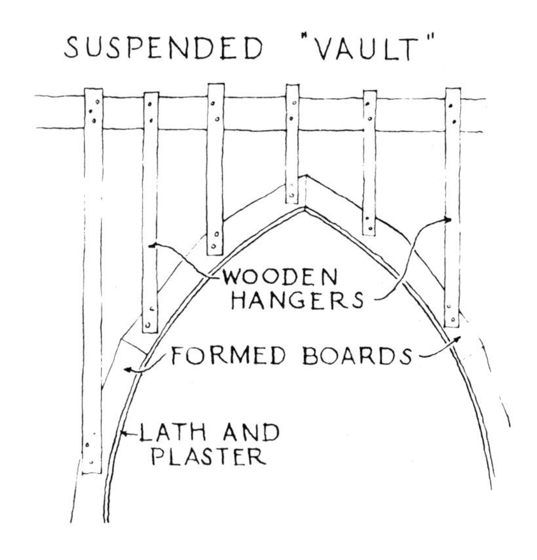

carried by projecting flagstones embedded in the wall. Triglyphs and moldings were of stucco. The contract, dated October 10, 1839, appears to have been carefully followed.

... columns as per plan, the flutes to be work'd in the brick work....

Cement The out side of walls columns, antaes, cornices and pediments to be cemented with best roman cement not air slack'd, prepared with coarse sand and fresh water gravel, the joints of the bricks to be well clean'd from loose mortar and the walls made sufficiently wet before the plastering is put on.

Refined exterior detailing in brick, covered with a thin coating of plaster. Beth Elohim Synagogue (1839), Charleston, S.C. (Louis Schwartz for Historic American Buildings Survey.)

the plastering of cement shall neatly develope the fluting of the columns of the portico and be finished in the best possible manner, all outside cementing must be done between the first of May and the first of August.[12]

Finishing. Exterior plaster was often scored to imitate the jointing of ashlar. These lines were sometimes marked with graphite or some other coloring to make them show up better. The color of the plaster was normally obtained from the sand and lime or natural cement, but brick dust or mineral pigments were sometimes added. Painted "joint" lines imitative of brickwork were occasionally drawn on the plaster; this process was called *bricking*.

ORNAMENTAL PLASTERWORK

Three kinds of special plasterwork are mentioned here: curved ceilings, plain and enriched moldings and conventional ornament in low relief.

Pseudo Vaults. Curved ceilings include the false vaults and domes of the 18th and 19th-century churches and public buildings. The desired curvature was obtained with formed boards covered with wood lath; these boards were suspended on wooden hangers fastened to roof trusses. In the late 19th century, metal lath came to be preferred for suspended ceilings, and with the advent of steel trusses, wires or small steel rods were used for hangers.

The auditorium of Beth Elohim Synagogue has a suspended ceiling in the form of a low dome. The dome and the arches beneath it are ornamented with coffers (deep panels). A specification written for the construction of the synagogue in 1839 describes 19th-century building practice that differed little from that of a hundred years earlier.

Plastering All the walls, partitions, ceilings etc. to be lath'd with saw'd laths four feet long to be put up with five nails to the lath, to have three coats plastering of strong lime sand and hair mortar, the third coat to be a hard finish, The ceiling to be as per plan. . . .

Dome & Ceiling To be furred down to receive plank ribbs and furred for five nails to the four foot lath, and nail'd securely to ranging timbers, all as per plan: The pannels in the spandrels to be sunk six inches, the mouldings planted in to be a four Inch moulding to have a margin of two Inches from the moulding to the face of the rails round the pannels, the pannels in the dome to be sunk six Inches at the bottom or lower end of the pannels and four and a half Inches at the upper end. The Mouldings to be four Inches at the lower end of the pannels and diminished in the same ratio of the sinking on the pannels at the upper ends. The ceiling of each story of the vestibule to have a plain centre piece.[13]

Moldings. There were two general methods of making moldings. If unornamented, they could be *run* in place. For example, if a room cornice was being installed, wooden brackets approximating the profile of the cornice would be placed at intervals of about one foot along the perimeter of the ceiling. Lathing would then be nailed to the brackets. All moldings except small ones were made hollow, to avoid shrinkage of plaster. (Plaster shrinks when laid too thickly.) Plastering over the forms required the efforts of at least two workmen. Screeds were first applied to the walls and ceiling at the edges of the proposed cornice. These served as guides to the template or *mold*. The mold was a thin wooden piece cut to the exact reverse of the molding desired; sharp corners and thin projections were reinforced with brass or other

TABLE 11—NAMES GIVEN TO PLASTER COATS

TWO-COAT

On brick or stone (called *rendered and set*):
First coat: *Rendering*
Second coat: *Set* or *setting coat*

On lath (called *laid and set*):
First coat: *Laying*
Second coat: *Set* or *setting-coat*

THREE-COAT

On brick or stone:
First coat: *Roughing* or *roughing in*

On lath:
First coat: *Pricking-up*
(By about the middle of the 19th century, the term *scratch coat* was used in the United States to denote the first coat.)

On brick, stone or lath:
Second coat: *Floating* or *floating-coat*
(The modern term is *brown coat.*)

On brick, stone or lath:
Third coat (to receive paper): *Set, setting-coat* or *finishing-coat*

On brick, stone or lath:
Third coat (to be painted): *Troweled stucco*, best quality. *Bastard stucco*, medium quality. *Rough stucco*, roughest quality.

The operations performed by a plasterer were called *rendering, laying, floating, setting* and *stuccoing*.

sheet metal. For wide cornices, several molds were needed. About one part plaster of paris was mixed with two parts lime putty to a semifluid consistency. One workman applied this plaster with a trowel while the second workman moved the mold along the form to remove superfluous material and secure an exact profile. The first workman supplied fresh plaster to those parts that needed it and sprinkled water with a brush on the portions being worked to keep them from drying too fast. Moldings were usually run before the second coat of plaster was applied to the walls and ceiling.

Sometimes plain moldings were cast; enriched (ornamented) moldings were always cast. Plaster of paris or a mixture of plaster of paris and lime putty was put into wax or glue molds. After hardening, the cast pieces were removed from the mold, scraped, cleaned and patched as needed. They were then fixed in place with cement made from plaster of paris. Castings seldom exceeded one foot in length. They were measured and priced by the lineal or running foot. Dealers kept various sizes in stock but special moldings could be made to order. Some ornamental castings were imported from Europe but they were also made in the colonies. In Baltimore: "Local craftsmen were advertising that their plaster decorations for interiors were superior to imported items and that they could furnish, in 1797, all sorts of molded 'landscape tablets, vases, rich flower festoons, wheat, vine and ivy, eagles, Apollo and Lyre,' etc."[14]

Relief Ornament. By the time ornamental plasterwork became common in the United States, the old practice of modeling relief ornaments in wet material was superseded by casting. The old technique, which was widely used in Italy during the 16th century spread to northern Europe and Great Britain.[15] This technique was modified in England, where repetitive parts of a design were cast and non-repetitive parts were modeled in place in wet plaster. Small dies were sometimes employed to impress ornaments in the moist plaster.

About the middle of the 18th century, papier-mâché ornaments made in French and British factories began to be used on walls and ceilings. They were fastened in place by being nailed to wooden blocks built into masonry walls or inserted into the frame of wooden buildings. Papier-mâché ornaments and wooden moldings were frequently combined with plain plaster surfaces. Ornaments cast from various patented materials called *composition* or *putty* were also used to enrich plasterwork. All of these materials that had been introduced in the 18th century continued to be popular as interior decoration in Europe during the 19th century and are also found in the United States.

REPAIR AND RESTORATION

Deteriorated interior walls and ceilings with plain surfaces are usually patched or replaced with modern plaster. This is especially true when they are papered or painted. Such repair work has always been considered a part of normal maintenance, and walls and ceilings in many old buildings have been repaired several times. The maintenance of ornamental plasterwork, except for keeping it clean, requires the specialized knowledge of an art conservator.

Deteriorated sections of plain plaster moldings can be repaired by running, or casts can be made from undamaged portions to replace badly damaged sections. Enriched moldings can also be replaced in sections with casts. Some dealers can supply moldings and small ornamental parts of the more standard period designs.

In patching and replacing damaged areas of exterior stucco walls, it is necessary to use material that matches as closely as possible the original material in proportion of mix, texture, density and appearance. Ingredients of the original material may be studied in the manner described in the chapter on mortar. Hard portland cement stucco is unsatisfactory for patching a lime-sand stucco or a natural cement stucco. Feather edges between a patch and old work should be avoided. Damaged moldings can be repaired in the same manner as interior moldings, using material suitable for outdoor exposure.

Plaster ceiling, c. 1745; the ornament was probably cast in sections and attached. First floor, Philipse Manor, Yonkers, N.Y. (Jack E. Boucher for the Division of Historic Preservation, New York State Office of Parks and Recreation).

Ornamental plaster ceiling, brackets and wall arcades. Farmers' and Exchange Bank (1853-54), Charleston, S.C. Francis D. Lee, architect. (Louis Schwartz for Historic American Buildings Survey.)

Suspended pseudo vaulting. Congregational (now First Unitarian) Church (1836-38), New Bedford, Mass. (Ned Goode for Historic American Buildings Survey.)

Ornamental plaster cornice and paneling. All panels have some wooden corners; the wainscot is of wood. Parlor, Morse-Libby House (c. 1850), Portland, Me. Historic American Buildings Survey).

Split-faced and rock-faced granite ashlar, simple trim. Congregational (First Unitarian) Church (1836-38), New Bedford, Mass. Alexander Jackson Davis and Russell Warren, architects. (Ned Goode for Historic American Buildings Survey.)

FOOTNOTES FOR CHAPTER IV

1. Sidney Fiske Kimball, *Domestic Architecture of the American Colonies and of the Early Republic* (New York: Charles Scribner's Sons, 1922), p. 30.
2. Frank Cousins and Phil M. Riley, *The Colonial Architecture of Salem* (Boston: Little, Brown, and Company, 1919), pp. 39–40.
3. Bainbridge Bunting and John P. Conron, "The Architecture of Northern New Mexico," *New Mexico Architecture*, vol. 8, no. 9–10 (1966), p. 20.
4. Arthur Evans, *The Palace of Minos at Knossos, vol. 1 (London: MacMillan, 1921), pp. 528–32.*
5. Richard B. Woodbury and Aubrey S. Trik, *The Ruins of Zaculea, Guatemala*, vol. 1. (United Fruit Company, 1953), p. 29.
6. J. Frederick Kelly, *The Early Domestic Architecture of Connecticut* (New Haven, Conn.: Yale University Press, 1924), p. 160.
7. Andrew Jackson Downing, *The Architecture of Country Houses* (New York: D. Appleton and Co., 1853), p. 368.
8. Edward H. Knight, *Knight's American Mechanical Dictionary* (New York: Hurd and Houghton, 1877). See individual terms.
9. Peter Nicholson, *The New Practical Builder* (London: Thomas Kelly, 1823), pp. 373–74.
10. F. E. Kidder, *Building Construction and Superintendence*, 9th ed., rev. by Thomas Nolan (New York: Comstock, 1910), p. 784.
11. Downing, op. cit., pp. 64–65.
12. Transcribed by the author by courtesy of Thomas J. Tobias, a descendant of the builder, David Lopez.
13. Ibid.
14. Richard Howland et al., *The Architecture of Baltimore* (Baltimore, Md.: Johns Hopkins Press, 1953), p. 6.
15. Vasari described the Italian method. *See* G. Baldwin Brown, ed., *Vasari on Technique* (New York: Dover, 1960), pp. 170–71.

APPENDIX
SUGGESTIONS FOR FURTHER STUDY

General

Norman Davey, *A History of Building Materials* (London: Phoenix House, 1961). For study of English and historic European materials.

Martin S. Briggs, *A Short History of the Building Crafts* (Oxford: Clarendon Press, 1925). For study of European crafts.

Charles Singer et al, eds., *A History of Technology* (London: Oxford University Press, 1954-1958). The five volumes of this monumental work contain chapters on European building materials and techniques.

Peter Nicholson, *The New Practical Builder* (London: Thomas Kelly, 1823).

The Mechanic's Companion (Philadelphia: James Locken, 1832).

Edward Lomax and Thomas Gunyon, eds., *Encyclopedia of Architecture* (New York: Johnson, Fry & Co., n.d.). Apparently published near the middle of the 19th century, this two-volume work contains much information from earlier times.

Edward H. Knight, *Knight's American Mechanical Dictionary* (New York: Hurd and Houghton, 1877). See entries on stone working, brick machines, plastering.

F. E. Kidder, *Building Construction and Superintendence*, 9th ed., rev. by Thomas Nolan (New York: Comstock, 1910). A standard work on late 19th-century practice.

Orin M. Bullock, Jr., *The Restoration Manual* (Norwalk, Conn.: Silvermine Publishers, 1966).

Stone

Wm. J. Arkell, *Oxford Stone* (London: Faber & Faber, 1947). Good historical treatment of the subject.

George P. Merrill, *Stones for Building and Decoration*, 3rd ed. (New York: Wiley, 1910). Although devoted to modern practices, this book contains some historical information.

Oliver Bowles, *The Stone Industries* (New York: McGraw-Hill, 1934).

Edward B. Mathews, "An Account of the Character and Distribution of Maryland Building Stones together with a History of the Quarrying Industry," *Maryland Geological Survey*, vol. 2 (Baltimore: Johns Hopkins Press, 1898). Mathews has given much more historical information than is normally found in geological surveys, including a good bibliography of early publications.

Ralph W. Stone, "Building Stones of Pennsylvania," *Pennsylvania Geological Survey*, 4th ser., bulletin M 15 (Harrisburg, Pa.: Topographic and Geologic Survey, 1932).

Arthur W. Brayley, *History of the Granite Industry of New England* (Boston: National Association of Granite Industries of the United States, 1913). A rambling account by a man who was long engaged in the industry.

Albert D. Hager, "Economical Geology of Vermont," *Report on the Geology of Vermont* (Claremont, N. H.: 1861).

Edwin C. Eckel, *Building Stones and Clays* (New York: Wiley, 1912). Eckel includes good bibliographies.

C. H. Hitchcock, "Economic Geology," "Building Materials," *The Geology of New Hampshire* (Concord, N. H.: Edward A. Jenks, 1878).

F. Noel Taylor, *Masonry as Applied to Civil Engineering* (New York: Van Nostrand, 1915).

W. P. Trowbridge, "Stone Cutting," *School of Mines Quarterly*, June 1883. Trowbridge describes the methods of stonecutters working on buildings for Columbia University; such detailed descriptions are rare. This article was reprinted under the title "Practical Stone Cutting" in *Carpentry and Building*, vol. 5, no. 9, Sept. 1883.

Harley J. McKee, "Early Ways of Quarrying and Working Stone in the United States," *Bulletin of the Association for Preservation Technology*, vol. 3, no. 1, 1971.

Brick

G. C. Mars, ed., *Brickwork in Italy* (Chicago: American Face Brick Association, 1925).

Nathaniel Lloyd, *A History of English Brickwork* (London: Montgomery, 1925).

Heinrich Ries, *Clays of New York:* Bulletin of the New York State Museum, no. 35, vol. 7, June 1900 (Albany, N. Y.: 1900).

Herbert A. Claiborne, *Comments on Virginia Brickwork before 1800* (The Walpole Society, 1957).

A. Lawrence Kocher, "Early Building with Brick," *Antiques*, July 1957.

Mortar

Col. Joseph G. Totten, *Essays on Hydraulic and Common Mortars and on Lime-burning* (Philadelphia: Franklin Institute, 1838).

Quincy Adams Gillmore, *Practical Treatise on Limes, Hydraulic Cements, and Mortars*, 5th ed. (New York: Van Nostrand, 1874).

Heinrich Ries and Edwin C. Eckel, *Lime and Cement Industries of New York* (Albany, N. Y.: 1901).

Ornamental Plastering

George P. Bankart, *The Art of the Plasterer* (London: Batsford, 1909).

Margaret Jourdain, *English Decorative Plasterwork of the Renaissance* (New York: Scribners, 1926).

INDEX